Andrew Catterall • Lynn Henfield • Christine Horbury

ESSENTIALS

AQA
GCSE Biology
Revision Guide

Contents

Contents

Unit 3

N.B. The numbers in brackets correspond to the reference numbers on the AQA GCSE Biology specification.

How to Use This Guide

This revision guide has been written and developed to help you get the most out of your revision.

This guide covers both Foundation and Higher Tier content.

(HT) Content that will only be tested on the Higher Tier papers appears in a pale yellow tinted box labelled with the (HT) symbol.

- The **coloured page headers** clearly identify the separate units, so that you can revise for each one separately: Unit 1 is red; Unit 2 is purple, and Unit 3 is blue.
- The exam will include questions on **How Science Works**, so make sure you work through the How Science Works section in green at the front of this guide before each exam.
- There are two **summary pages** at the end of each unit, which outline all the key points. These are great for a final recap before your exam.

- There are **practice questions** at the end of each unit so you can test yourself on what you've just learned. (The answers are given on page 76 so you can mark your own answers.)
- You'll find **key words** in a yellow box on each 2-page spread. They are also highlighted in colour within the text; higher tier key words are highlighted in a different colour. Make sure you know and understand all these words before moving on!
- There's a **glossary** at the back of the book. It contains all the key words from throughout the book so you can check any definitions you're unsure of.
- The **tick boxes** on the contents page let you track your revision progress: simply put a tick in the box next to each topic when you're confident that you know it.
- Don't just read the guide — **learn actively**! Constantly test yourself without looking at the text.

Good luck with your exam!

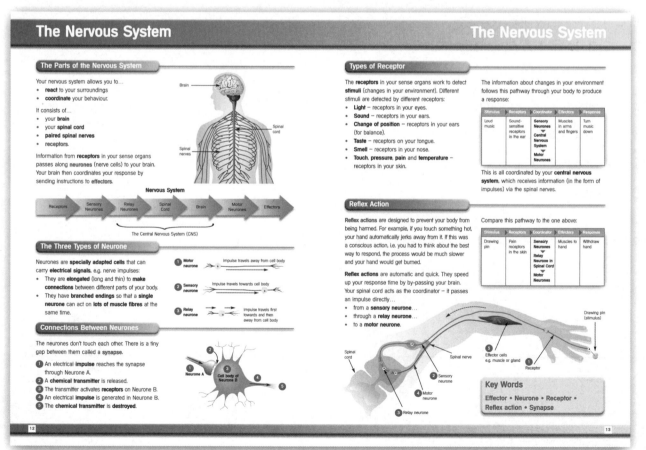

How Science Works – Explanation

The AQA GCSE science specifications incorporate:

1. **Science Content** – all the scientific explanations and evidence that you need to know for the exams. (It is covered on pages 12–71 of this revision guide.)

2. **How Science Works** – a set of key concepts, relevant to all areas of science. It covers…
   - the relationship between scientific evidence, and scientific explanations and theories
   - how scientific evidence is collected
   - how reliable and valid scientific evidence is
   - the role of science in society
   - the impact science has on our lives
   - how decisions are made about the ways science and technology are used in different situations, and the factors affecting these decisions.

Your teacher(s) will have taught these two types of content together in your science lessons. Likewise, the questions on your exam papers will probably combine elements from both types of content. So, to answer them, you'll need to recall the relevant scientific facts *and* apply your knowledge of how science works.

The key concepts of How Science Works are summarised in this section of the revision guide (pages 6–11).

You should be familiar with all of these concepts. If there is anything you are unsure about, ask your teacher to explain it to you.

How Science Works is designed to help you learn about and understand the practical side of science. It aims to help you develop your skills when it comes to…
- evaluating information
- developing arguments
- drawing conclusions.

N.B. Practical tips on how to evaluate information are included on page 11.

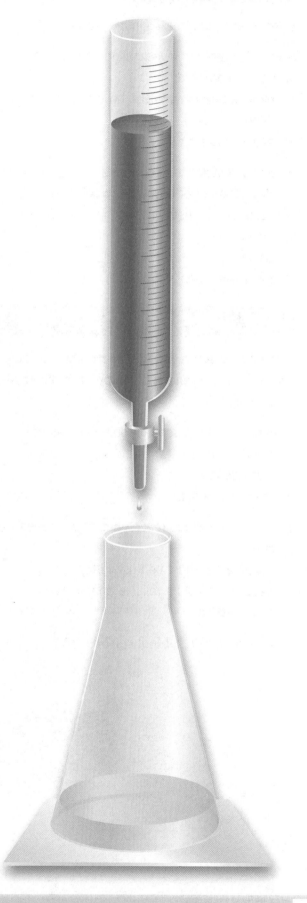

How Science Works

What is the Purpose of Science?

Science attempts to explain the world we live in.

Scientists carry out investigations and collect evidence in order to…

- **explain phenomena** (i.e. how and why things happen)
- **solve problems**.

Scientific knowledge and understanding can lead to the **development of new technologies** (e.g. in medicine and industry), which have a huge impact on…

- society
- the environment.

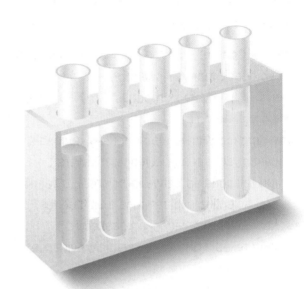

What is the Purpose of Evidence?

Scientific evidence provides **facts** which answer a specific question and either **support** or **disprove** an idea / theory.

Evidence is often based on data that has been collected through…

- **observations**
- **measurements**.

To allow scientists to reach conclusions, evidence must be…

- **reliable** – it must be trustworthy
- **valid** – it must be reliable and answer the question.

N.B. If data isn't reliable, it can't be valid.

To ensure scientific evidence is reliable and valid, scientists use ideas and practices relating to…

1. observations
2. investigations
3. measurements
4. data presentation
5. conclusions.

These five key ideas are covered in more detail on the following pages.

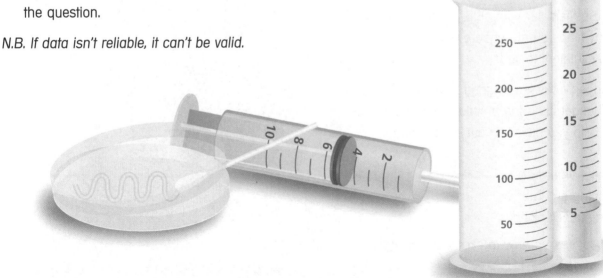

Observations

Most scientific investigations begin with an **observation**. A scientist observes an event / phenomenon and decides to find out more about how and why it happens.

The first step is to develop a **hypothesis** to suggest an explanation for the phenomenon. Hypotheses normally suggest a relationship between two or more **variables** (factors that change). Hypotheses are based on…

- careful observations
- existing scientific knowledge
- a bit of creative thinking.

The hypothesis is used to make a **prediction**, which can be tested through scientific investigation. The data collected from the investigation will…

- support the hypothesis **or**
- show it to be untrue **or**
- lead to the development of a new hypothesis.

If new observations or data don't match existing explanations or theories, they must be checked for reliability and validity.

Sometimes, if the new observations and data are valid, existing theories and explanations have to be revised or amended, and so scientific knowledge grows and develops.

Example

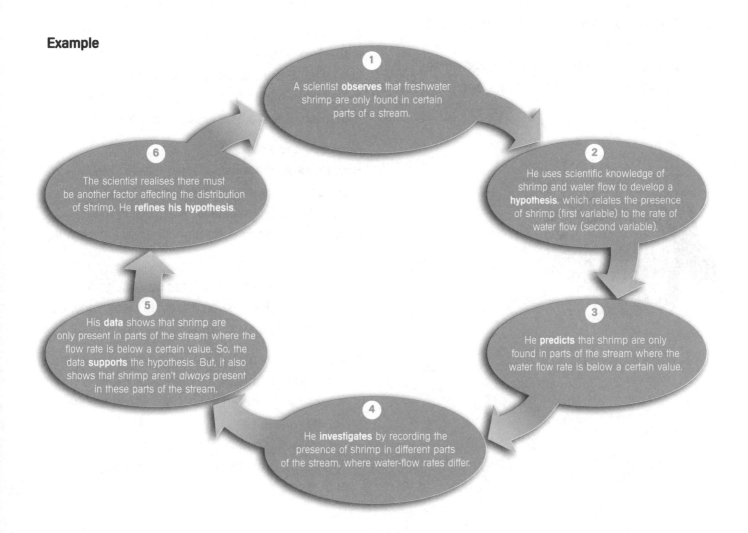

1. A scientist **observes** that freshwater shrimp are only found in certain parts of a stream.

2. He uses scientific knowledge of shrimp and water flow to develop a **hypothesis**, which relates the presence of shrimp (first variable) to the rate of water flow (second variable).

3. He **predicts** that shrimp are only found in parts of the stream where the water flow rate is below a certain value.

4. He **investigates** by recording the presence of shrimp in different parts of the stream, where water-flow rates differ.

5. His **data** shows that shrimp are only present in parts of the stream where the flow rate is below a certain value. So, the data **supports** the hypothesis. But, it also shows that shrimp aren't *always* present in these parts of the stream.

6. The scientist realises there must be another factor affecting the distribution of shrimp. He **refines his hypothesis**.

How Science Works

An **investigation** involves collecting data to find out whether there is a relationship between two **variables**. A variable is a factor that can take different values.

In an investigation there are two variables:

1. **Independent variable** – can be adjusted (changed) by the person carrying out the investigation.
2. **Dependent variable** – measured each time a change is made to the independent variable, to see if it also changes.

Variables can have different types of values:

- **Continuous variables** – take numerical values. These are usually measurements, e.g. temperature.
- **Discrete variables** – only take whole-number values. These are usually quantities, e.g. the number of shrimp in a stream.
- **Ordered variables** – have relative values, e.g. 'small', 'medium' or 'large'.
- **Categoric variables** – have a limited number of specific values, e.g. different breeds of dog.

N.B. Numerical values tend to be more informative than ordered and categoric variables.

An investigation tries to find out whether an **observed** link between two variables is…

- **causal** – a change in one variable causes a change in the other
- **due to association** – the changes in the two variables are linked by a third variable
- **due to chance** – the change in the two variables is unrelated; it is coincidental.

Fair Tests

In a **fair test**, the only factor that can affect the dependent variable is the independent variable. Other **outside variables** that could influence the results are kept the same or eliminated.

It's a lot easier to carry out a fair test in the lab than in the field, where conditions can't always be controlled. The impact of an outside variable, like the weather, has to be reduced by ensuring all measurements are affected by it in the same way.

Accuracy and Precision

How accurate the data collected needs to be depends on what the investigation is trying to find out. For example, measures of alcohol in the blood must be accurate to determine whether a person is legally fit to drive.

The data collected must be **precise** enough to form a **valid conclusion**: it should provide clear evidence for or against the hypothesis.

To ensure data is as accurate as possible, you can…

- calculate the **mean** (average) of a set of repeated measurements to get a **best estimate** of the true value
- increase the number of measurements taken to improve the **accuracy** and the **reliability** of the mean.

Measurements

Apart from outside variables, there are a number of factors that can affect the reliability and validity of measurements:

- **Accuracy of instruments** – depends on how accurately the instrument has been calibrated. (Expensive equipment is usually more accurately calibrated.)
- **Sensitivity of instruments** – determined by the smallest change in value that the instrument can detect. For example, bathroom scales aren't sensitive enough to detect the changes in a baby's weight, but the scales used by a midwife are.

- **Human error** – can occur if you lose concentration. Systematic (repeated) errors can occur if the instrument hasn't been calibrated properly or is misused.

You need to examine any **anomalous** (irregular) values to try to determine why they appear. If they have been caused by an equipment failure or human error, it is common practice to discount them from any calculations.

Presenting Data

Data is often presented in a **chart** or **graph** because it makes…

- the patterns more evident
- it easier to see the relationship between two variables.

The relationship between variables can be…

- **linear** (positive or negative), **or**
- **directly proportional**.

If you present data clearly, it is easier to identify any anomalous values. The type of chart or graph you use to present data depends on the type of variable involved:

1. **Tables** organise data (but patterns and anomalies aren't always obvious).
2. **Bar charts** display data when the independent variable is categoric or discrete and the dependent variable is continuous.
3. **Line graphs** display data when both variables are continuous.
4. **Scattergrams** (scatter diagrams) show the underlying relationship between two variables. This can be made clearer if you include a **line of best fit**.

1

Height (cm)	127	165	149	147	155	161	154	138	145
Shoe Size	5	8	5	6	5	5	6	4	5

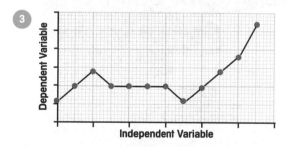

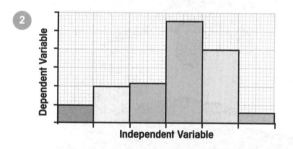

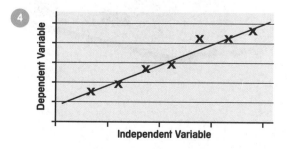

Conclusions

Conclusions **should**…
- describe the patterns and relationships between variables
- take all the data into account
- make direct reference to the original hypothesis / prediction.

Conclusions **shouldn't**…
- be influenced by anything other than the data collected
- disregard any data (except anomalous values)
- include any speculation.

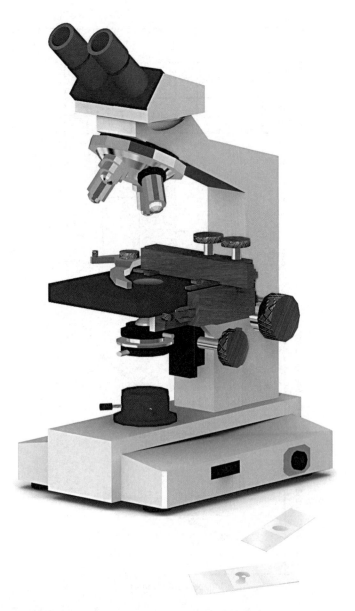

Evaluation

An **evaluation** looks at the whole investigation. It should consider…
- the original purpose of the investigation
- the appropriateness of the methods and techniques used
- the reliability and validity of the data
- the validity of the conclusions.

The **reliability** of an investigation can be increased by…
- looking at relevant data from secondary sources
- using an alternative method to check results
- ensuring that the results can be reproduced by others.

Science and Society

Scientific understanding can lead to technological developments. These developments can be exploited by different groups of people for different reasons. For example, the successful development of a new drug…
- benefits the drugs company financially
- improves the quality of life for patients.

Scientific developments can raise certain **issues**. An issue is an important question that is in dispute and needs to be settled. The resolution of an issue may not be based on scientific evidence alone.

There are several different **issues** which can arise:
- **Social** – the impact on the human population of a community, city, country, or the world.
- **Economic** – money and related factors like employment and the distribution of resources.
- **Environmental** – the impact on the planet, its natural ecosystems and resources.
- **Ethical** – what is morally right and wrong; requires a valued judgement to be made about what is acceptable.

N.B. There is often an overlap between social and economic issues.

Evaluating Information

It is important to be able to evaluate information relating to social-scientific issues, both for the exam and to help you make informed decisions in life.

When evaluating information…
- make a list of **pluses**
- make a list of **minuses**
- consider how each point might **impact on society**.

*N.B. Remember, **PMI** – **p**luses, **m**inuses, **i**mpact on society.*

You also need to consider whether the source of information is reliable and credible. Some important factors to consider are…
- **opinions**
- **bias**
- **weight of evidence**.

Opinions are personal viewpoints. Opinions backed up by valid and reliable evidence carry far more weight than those based on non-scientific ideas.

Information is **biased** if it favours one particular viewpoint without providing a balanced account. Biased information might include incomplete evidence or try to influence how you interpret the evidence.

Scientific evidence can be given **undue weight** or dismissed too lightly due to…
- political significance
- status (academic or professional status, experience, authority and reputation).

Limitations of Science

Although science can help us in lots of ways, it can't supply all the answers. We are still finding out about things and developing our scientific knowledge.

There are some questions that science can't answer. These tend to be questions relating to…
- ethical issues
- situations where it isn't possible to collect reliable and valid scientific evidence.

Science can often tell us if something **can** be done, and **how** it should be done, but it can't tell us whether it **should** be done.

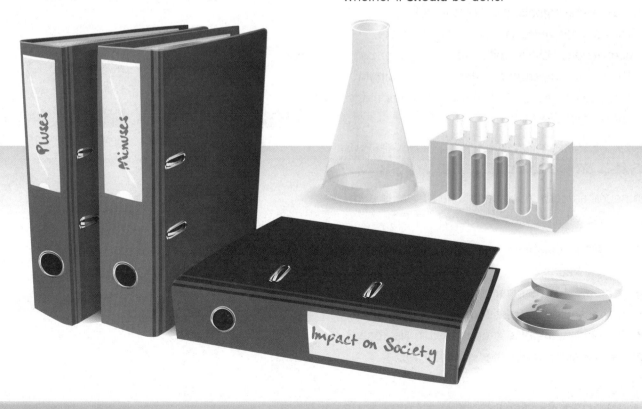

The Nervous System

The Parts of the Nervous System

Your nervous system allows you to...
- **react** to your surroundings
- **coordinate** your behaviour.

It consists of...
- your **brain**
- your **spinal cord**
- **paired spinal nerves**
- **receptors**.

Information from **receptors** in your sense organs passes along **neurones** (nerve cells) to your brain. Your brain then coordinates your response by sending instructions to **effectors**.

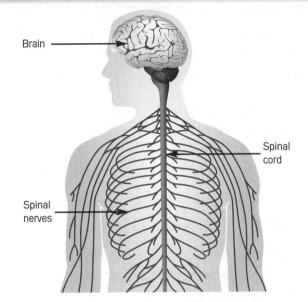

Brain

Spinal cord

Spinal nerves

Nervous System

Receptors → Sensory Neurones → Relay Neurones → Spinal Cord → Brain → Motor Neurones → Effectors

The Central Nervous System (CNS)

The Three Types of Neurone

Neurones are **specially adapted cells** that can carry **electrical signals**, e.g. nerve impulses:
- They are **elongated** (long and thin) to **make connections** between different parts of your body.
- They have **branched endings** so that a **single neurone** can act on **lots of muscle fibres** at the same time.

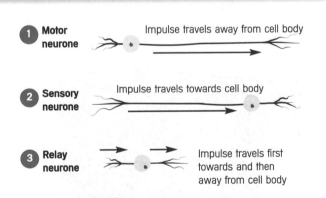

1. **Motor neurone** — Impulse travels away from cell body
2. **Sensory neurone** — Impulse travels towards cell body
3. **Relay neurone** — Impulse travels first towards and then away from cell body

Connections Between Neurones

The neurones don't touch each other. There is a tiny gap between them called a **synapse**.

1. An electrical **impulse** reaches the synapse through Neurone A.
2. A **chemical transmitter** is released.
3. The transmitter activates **receptors** on Neurone B.
4. An electrical **impulse** is generated in Neurone B.
5. The **chemical transmitter** is **destroyed**.

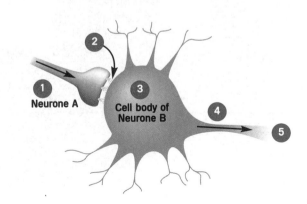

Neurone A

Cell body of Neurone B

Types of Receptor

The **receptors** in your sense organs work to detect **stimuli** (changes in your environment). Different stimuli are detected by different receptors:

- **Light** – receptors in your eyes.
- **Sound** – receptors in your ears.
- **Change of position** – receptors in your ears (for balance).
- **Taste** – receptors on your tongue.
- **Smell** – receptors in your nose.
- **Touch, pressure, pain** and **temperature** – receptors in your skin.

The information about changes in your environment follows this pathway through your body to produce a response:

Stimulus	Receptors	Coordinator	Effectors	Response
Loud music	Sound-sensitive receptors in the ear	**Sensory Neurones** ▼ **Central Nervous System** ▼ **Motor Neurones**	Muscles in arms and fingers	Turn music down

This is all coordinated by your **central nervous system**, which receives information (in the form of impulses) via the spinal nerves.

Reflex Action

Reflex actions are designed to prevent your body from being harmed. For example, if you touch something hot, your hand automatically jerks away from it. If this was a conscious action, i.e. you had to think about the best way to respond, the process would be much slower and your hand would get burned.

Reflex actions are automatic and quick. They speed up your response time by by-passing your brain. Your spinal cord acts as the coordinator – it passes an impulse directly…

- from a **sensory neurone**…
- through a **relay neurone**…
- to a **motor neurone**.

Compare this pathway to the one above:

Stimulus	Receptors	Coordinator	Effectors	Response
Drawing pin	Pain receptors in the skin	**Sensory Neurones** ▼ **Relay Neurone in Spinal Cord** ▼ **Motor Neurones**	Muscles to hand	Withdraw hand

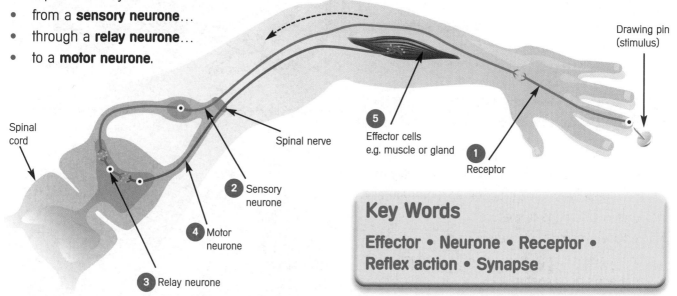

Spinal cord

Spinal nerve

Drawing pin (stimulus)

5 Effector cells e.g. muscle or gland

1 Receptor

2 Sensory neurone

4 Motor neurone

3 Relay neurone

Key Words

Effector • Neurone • Receptor • Reflex action • Synapse

Internal Environment and Hormones

Internal Conditions

Humans need to keep their internal environment relatively constant. This means your body must control…

- **temperature** – to maintain the temperature at which most body enzymes work best (37°C)
- **blood sugar (glucose) levels** – to constantly supply cells with energy
- **water content**
- **ion content**.

Your body raises its temperature by shivering, and lowers it by sweating.

You gain glucose, water and ions by eating and drinking.

Glucose is used up as energy when you move. Water and ions are lost by sweating and through your kidneys in urine. Water is also lost through your lungs when you breathe.

Hormones

Many processes within your body are coordinated by **hormones**. Hormones are produced by **glands**, and transported to their target organs by the bloodstream.

Hormones regulate the functions of many organs and cells.

Women produce hormones that cause…
- eggs to be released from the **ovaries**
- changes in the thickness of the **womb** lining.

These hormones are produced by the **pituitary gland** and the ovaries.

Natural Control of Fertility

1 **FSH** from the pituitary gland causes…
  - the ovaries to produce oestrogen
  - an egg to mature.

2 **Oestrogen** from the ovaries…
  - slows down the production of FSH
  - causes the production of LH.

3 **LH** from the pituitary gland…
  - stimulates the release of an egg in the middle of the **menstrual cycle**.

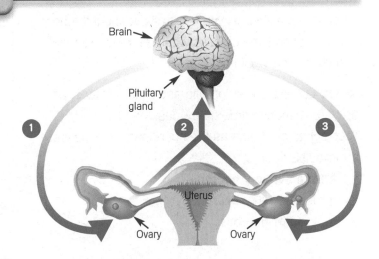

Artificial Control of Fertility

FSH and oestrogen can be given to women to…
- **increase fertility**: FSH is given as a fertility drug to women who don't produce enough naturally to stimulate eggs to mature
- **reduce fertility**: oestrogen is given as an oral contraceptive to slow down FSH production so that eggs don't mature.

Metabolic Rate

Metabolic rate is the rate at which all **chemical reactions** in your body's cells are carried out. It is affected by...

- the amount of **activity / exercise** you do
- the **proportion of fat to muscle** in your body
- **inherited** factors.

If you exercise regularly, you are likely to be fitter than people who don't. Your metabolic rate will stay high for some time after you have finished exercising. The less exercise you take and the warmer it is, the less food you need.

Healthy Diets

A **healthy diet** contains the right balance of the different foods your body needs (i.e. carbohydrates, fat, protein, fibre, vitamins, minerals and water). A person is **malnourished** if their diet isn't balanced. A poor diet can lead to...

- a person being **too fat** or **too thin**
- **deficiency diseases** such as **scurvy** and **rickets**.

In the **developed** world (e.g. UK, USA) people often eat too much food and don't do enough exercise. This is leading to high levels of **obesity** and conditions like...

- arthritis
- diabetes
- high blood pressure
- heart disease.

In the **developing** world (e.g. parts of Africa and Asia) people suffer from problems due to lack of food, like...

- reduced resistance to infection
- irregular periods in women.

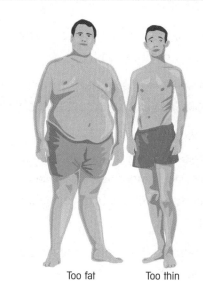

Too fat   Too thin

Key Words

Deficiency disease • FSH • Gland • Hormone • Ion • LH • Malnourished • Menstrual cycle • Metabolic rate • Obesity • Pituitary gland

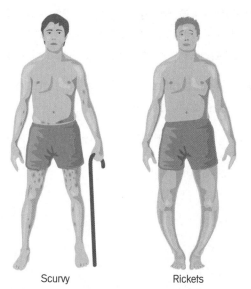

Scurvy   Rickets

Diet

Cholesterol

Cholesterol is made in your **liver** and is found in your **blood**.

The amount of cholesterol your liver makes depends on...
- your diet
- inherited factors.

Cholesterol is carried around your body by two types of **lipoproteins**:
- **Low-density lipoproteins** (**LDLs**) are 'bad' cholesterols — high levels in the blood can cause disease of the heart and blood vessels.
- **High-density lipoproteins** (**HDLs**) are 'good' cholesterols.

The balance of these types of cholesterol is important for a healthy heart.

Ideal Amounts of Cholesterol in the Blood

LDLs	**HDLs**
Cause heart disease	Good health

Fats and Salt

There are two types of fats:
- **Saturated fats** (e.g. butter, animal fat).
- **Unsaturated** (**monounsaturated and polyunsaturated**) **fats** (e.g. vegetable oil, some fish oil).

Saturated fats are 'bad' fats — they increase blood sugar levels.

Unsaturated fats are 'good' fats — they help to reduce blood cholesterol levels and improve the balance between 'good' and 'bad' lipoproteins.

Processed food often contains a high proportion of saturated fat and / or salt.

Too much **salt** in your diet can lead to increased blood pressure.

30% of people in the UK consume too much salt

Drugs

Drugs are chemical substances which alter the way your body works. They can be **beneficial** but they can also **harm** your body.

- Some drugs can be obtained from **natural** substances (many have been known to indigenous people for years).
- Some drugs are **synthetic** (developed by scientists).

Developing New Drugs

When new **medical drugs** are developed, they need to be thoroughly **tested** and **trialled** in the laboratory to find out if they are **toxic** (poisonous). Then they are checked for **side-effects** on human volunteers.

The flow chart shows the stages in developing a new drug.

New drug made in laboratory → Tested in laboratory for toxicity → Trialled on volunteers to check for side-effects

Thalidomide

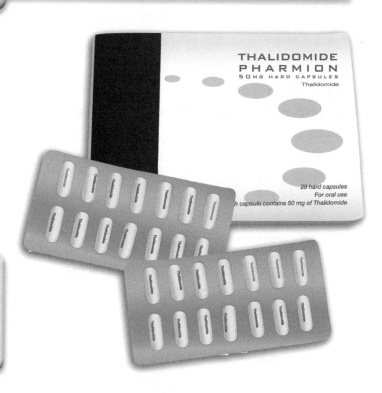

- The **thalidomide** drug was developed, tested and **approved** as a sleeping pill.
- It was also found to be effective in relieving morning sickness in pregnant women. But it **hadn't been tested** for this use.
- It was **banned** when many women who took the drug gave birth to babies with severe limb abnormalities.
- Thalidomide was re-tested.
- It is now used to treat **leprosy**.

Key Words

Cholesterol • Drug • Leprosy • Lipoproteins • Saturated fat • Side-effect • Unsaturated fat

Drugs

Legal and Illegal Drugs

Some people use drugs illegally for **recreation** (pleasure), but they can be very harmful. Some recreational drugs are **legal** (alcohol, tobacco); some are **illegal** (e.g. heroin and cocaine, which are very addictive).

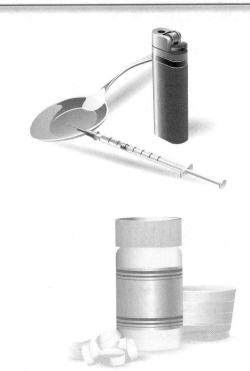

Legal drugs have a bigger overall impact on health than illegal drugs, because far more people use them.

- Drugs alter chemical processes in the body, so people can become dependent on, or **addicted** to, them.
- Once they are addicted, people will suffer **withdrawal symptoms** if they don't have the drugs.
- Withdrawal symptoms may be **psychological** and **physical** (e.g. paranoia, sweating, vomiting).

Alcohol and Tobacco

Alcohol contains the chemical **ethanol**. Alcohol…
- affects the nervous system, causing reactions to slow down
- helps people to relax
- excess can lead to a lack of self-control, unconsciousness, coma or death
- can lead to liver damage or brain damage in the long-term.

Tobacco smoke contains carbon monoxide, nicotine (which is addictive) and **carcinogens**. Tobacco can cause…
- emphysema – alveoli damage caused by coughing
- bronchitis
- arterial and heart disease
- lung cancer.

The carbon monoxide in tobacco smoke reduces the oxygen-carrying capacity of the blood. In pregnant women this can…
- deprive the **fetus** of oxygen
- lead to a **low birth mass**.

Key Words

Antibiotics • Carcinogen • Fetus • Infectious • MRSA • Natural selection • Pathogen • Toxin • Vaccine

Pathogens

Microorganisms that cause **infectious** diseases are called **pathogens**. There are two main types of pathogen that can affect your health:

- **bacteria**
- **viruses**.

White blood cells help to defend against pathogens by...

- ingesting pathogens
- producing **antitoxins** to neutralise **toxins** from the pathogens
- producing **antibodies** to destroy certain pathogens.

Bacteria	Viruses
Very small.	Smaller than bacteria.
Reproduce rapidly.	Reproduce rapidly once inside living cells, which they then damage.
Can produce toxins which make you feel ill.	Can produce toxins which make you feel ill.
Cause illnesses like TB, tetanus, cholera.	Cause illnesses like colds, flu, measles, polio.

Treatment of Disease

Medicines such as **painkillers** (e.g. aspirin) are used to **alleviate the symptoms** of disease. However, painkillers don't kill pathogens.

Antibiotics (e.g. penicillin) are used to **kill the infective bacterial pathogens** inside the body. However, antibiotics **don't kill viral pathogens** that live and reproduce inside cells.

It's difficult to develop drugs that kill viruses without damaging your body's tissues.

Many strains of bacteria, including **MRSA**, have developed **resistance** to antibiotics as a result of **natural selection**. So, it's necessary to prevent over-use of antibiotics.

Vaccination

You can acquire immunity to a particular disease by being **vaccinated (immunised)**.

1. An inactive / dead pathogen is injected into your body.
2. Your white blood cells produce antibodies to destroy the pathogen.
3. You then have an **acquired immunity** to this particular pathogen because your white blood cells are sensitised to it and will respond to any future infection by producing antibodies quickly.

An example of a **vaccine** is the MMR vaccine used to protect children against measles, mumps and rubella.

Unit 1a Summary

The Nervous System

Stimulus ➡ Receptor ➡ Coordinator ➡ Effector ➡ Response

Conscious Action:

Central nervous system (coordinator) = Brain + Spinal cord + Relay neurones

Receptor ➡ Sensory neurone ➡ CNS ➡ Motor neurone ➡ Effector

Reflex Action:

- Coordinator = Spinal cord
- Pathway by-passes brain to produce quick, automatic response.

Receptor ➡ Sensory neurone ➡ Relay neurone in spinal cord ➡ Motor neurone ➡ Effector

The gaps between connecting neurones are called **synapses**. These are bridged through the production of a chemical transmitter.

Internal Environment

Conditions that must be controlled = Temperature + Blood glucose + Water content + Ion content

Hormones and Fertility

- Hormones regulate the functions of many organs and cells.
- Fertility in women is controlled by hormones.

Hormone	Source	Function
FSH	Pituitary gland	• Causes egg to mature. • Causes ovaries to produce oestrogen.
Oestrogen	Ovaries	• Slows down FSH production. • Causes LH production.
LH	Pituitary gland	• Stimulates release of egg.

- FSH can be used as a fertility drug – stimulates eggs to mature.
- Oestrogen can be used as an oral contraceptive – inhibits FSH so no eggs mature.

Metabolism and Diet

- Metabolic rate = Rate at which chemical reactions in the body's cells take place.
- Metabolism is affected by exercise, ratio of fat to muscle, and genes.
- Healthy diet = Correct balance of different food groups (dependent on individual).
- A poor diet can lead to malnourishment and **deficiency diseases**, e.g. scurvy and rickets.

Cholesterol, Fats and Salt

Cholesterol is made in liver; dependent on diet and genes; carried by LDLs (bad) and HDLs (good).

Type of Fat	Examples	Good / Bad	Why?
Saturated	Butter, animal fat	Bad	Increases blood glucose.
Unsaturated	Vegetable oil, some fish oil	Good	Reduces cholesterol.

Too much salt ➡ High blood pressure

Drugs

Drugs = Chemical substances, which alter the way the body works.

- Drugs can be synthetic or natural.
- New medical drugs must be tested in laboratory and trialled on volunteers for safety and effectiveness.
- Many drugs are addictive.
- Addicts suffer withdrawal symptoms when they stop using them – psychological and physical.

Alcohol and Tobacco

Alcohol...
- contains the chemical ethanol
- slows the nervous system
- can damage liver and brain.

Tobacco smoke...
- contains carbon monoxide, nicotine and cancer-causing **carcinogens**
- can cause emphysema, bronchitis, heart disease and lung cancer
- can lead to low birth mass in babies born to smokers.

Pathogens

Pathogens = Bacteria and viruses.

White blood cells defend against pathogens by ingesting them, and by producing **antitoxins** and **antibodies**.

Painkillers alleviate symptoms but don't kill pathogens. Antibiotics kill bacterial pathogens but don't kill viral pathogens.

Vaccination provides immunity to a particular disease:

Dead / inactive pathogen injected ➡ Antibodies produced by white blood cells ➡ Acquired immunity

Unit 1a Practice Questions

1 Match the words A, B, C and D with the parts labelled 1 to 4 in the diagram below.

A receptor ....

B effector ....

C stimulus ....

D spinal cord ....

2 What is the name of the tiny gap between neurones? ....

3 Match the words A, B, C and D with the parts labelled 1 to 4 in the diagram below.

A pituitary gland ....

B ovary ....

C uterus ....

D brain ....

4 a) FSH can be given to women as a fertility drug. Explain how FSH can increase a woman's fertility.

....

....

b) Which hormone is made and released by the ovaries? ....

c) Explain why the hormone oestrogen is used as an oral contraceptive.

....

d) What is the name of the hormone which causes the eggs to be released? ....

5 Match the words A, B, C and D with the spaces numbered 1 to 4 in the sentences below.

A malnourished ....   **B** obese ....

C carbohydrates ....   **D** scurvy ....

To ensure you have a balanced diet you need to control your intake of protein, __1__ , fats and fibre. A person is __2__ if their diet isn't balanced. Deficiency diseases like __3__ are caused by a lack of vitamins. People who consume too much food and do too little exercise become __4__ .

6 What is meant by the term 'metabolic rate'?

....

....

7 Match the drugs A, B, C and D with the descriptions numbered 1 to 4 below.

A heroin ....

B tobacco ....

C alcohol ....

D illegal drugs ....

1 Causes reactions to slow down.

2 Contains carcinogens.

3 Very addictive.

4 Used by some people for recreation.

8 Match the words A, B, C and D with the spaces numbered 1 to 4 in the sentences below.

A painkillers ....

B antibiotics ....

C pathogens ....

D white blood cells ....

Microorganisms that cause disease are called ___1___ . ___2___ often alleviate the symptoms of disease but don't kill the pathogen. The body's immune system uses ___3___ to fight infections caused by pathogens. To kill bacteria that cause infections we use ___4___ .

9 This question has three parts. For each part, put a tick next to the correct answer.

a) Which of these sequences describes how immunisation works?

i) Dead pathogen injected – White blood cells sensitised – Antibodies produced – Immunity.

ii) Antibodies produced – Dead pathogen injected – Immunity – White blood cells sensitised.

iii) White blood cells sensitised – Dead pathogen injected – Antibodies produced – Immunity.

iv) Dead pathogen injected – Antibodies produced – White blood cells sensitised – Immunity.

b) Which of these statements is true?

i) Antibiotics can be used to fight bacteria and viruses.

ii) Antibiotics can only fight infections caused by viruses.

iii) Antibiotics are produced by white blood cells.

iv) Antibiotics can be used to fight infections caused by bacteria.

c) White blood cells form part of the body's immune system. Which of these statements is false?

i) White blood cells produce antitoxins.

ii) White blood cells ingest pathogens.

iii) White blood cells produce antibiotics.

iv) White blood cells produce antibodies.

10 Why is it necessary to prevent overuse of antibiotics?

....

Competition and Adaptations

Populations and Communities

A **population** is the total number of individuals of the same species which live in a certain area.

A **community** is the total number of organisms in a particular area.

Competition

In order to **survive**, organisms need materials from their surroundings, and from the other living organisms there.

Organisms **compete** with each other for **space / light, food** and **water**.

Plants compete with each other for…
- light
- water
- nutrients from the soil.

Animals compete with each other for…
- food
- water
- mates
- territory.

When organisms compete, those which are **better adapted** to their environment are more successful and usually exist in larger numbers. This often leads to the exclusion of other competing organisms.

Adaptations

Adaptations are features that make an organism well-suited to its environment.

Adaptations increase an organism's chance of **survival**.

Animals and plants may be adapted for survival in the conditions of their environment. For example, the polar bear is well-suited to its habitat in the Arctic:
- it has a white coat so it is camouflaged
- it has a lot of insulating fat beneath its skin
- it has large feet to spread its weight on the ice.

Animals and plants may be adapted to cope with particular features of their environment. For example…
- some plants (e.g. cacti) have thorns / spines to prevent animals eating them
- some animals (e.g. blue dart frogs) have developed poisons and warning colours to keep **predators** away.

Genetic Information

- The nucleus of a cell contains **chromosomes**.
- **Chromosomes** are made up of a substance called **DNA**.
- A section of a chromosome is called a **gene**.
- Different **genes** control the development of different **characteristics**.

During **reproduction**, genes are passed from parent to offspring (they are **inherited**).

Chromosomes come in **pairs**, but different species have different numbers of pairs. Humans have **23 pairs**.

Key Words

Adaptation • Asexual reproduction • Chromosome • Clone • DNA • Gamete • Gene • Predator • Sexual reproduction • Variation

A Section of One Chromosome

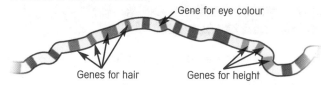

Gene for eye colour

Genes for hair

Genes for height

Variation

Differences between individuals of the same species are called **variation**. Variation may be due to **genetic** or **environmental factors**.
- **Genetic** factors are responsible for things such as the colour of dogs' coats being different.
- **Environmental** factors are responsible for identical twins being very different in weight. If one twin has a diet high in fat and doesn't exercise, he will become fatter than his brother.

Reproduction and Variation

During **sexual reproduction** male and female **gametes** fuse together. The genes carried by the egg and the sperm are mixed together to produce a new individual. This process produces lots of variation, even amongst offspring from the same parents.

Sexual reproduction 　Variation

Asexual reproduction doesn't produce any variation at all, unless it is due to environmental causes. Only one parent is needed to produce individuals who are genetically identical to the parent, i.e. **clones**.

Asexual reproduction 　No variation

Genetics

Reproducing Plants

Plants can reproduce **asexually**. Offspring produced asexually are **clones**. Many plants naturally reproduce asexually (e.g. spider plants).

Many plants can be reproduced asexually **by artificial means**. For example, you can take **cuttings** from a plant with desired characteristics to produce clones quickly and cheaply.

Spider Plant Stolons

Stolon – a rooting side branch | New individual established | Now independent

Plant Cuttings

Cloning Techniques

These are some modern cloning techniques:

1 **Tissue culture** – small groups of cells are scraped from part of a plant and grown in soil containing nutrients and hormones. The offspring are **genetically identical to the parent plant**.

2 **Embryo transplants** – cells from a developing animal are split before they become **specialised**. These identical embryos are transplanted into host mothers. The offspring are **genetically identical to each other, but not to the parents**.

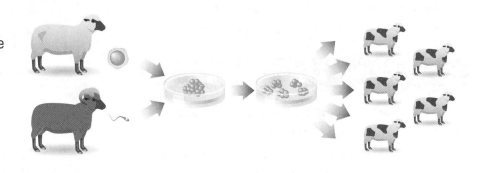

3 **Adult cell / fusion cell cloning** – DNA from a donor animal is inserted into an empty egg cell. This develops into an embryo and is implanted into another female. The female produces offspring that are **genetically identical to the donor animal**.

Genetic Modification

Genetic modification (**genetic engineering**) involves transferring genetic material from one organism to another.

Genes from the chromosomes of humans and other organisms are **cut out** using **enzymes**. Then they are transferred to cells of other organisms.

Reasons for altering an organism's genetic make-up are…

- to improve the crop yield
- to improve resistance to pests or herbicides
- to extend the shelf-life of fast-ripening crops
- to harness the cell chemistry of an organism so that it produces a required substance, e.g. production of human **insulin**.

In animals and plants, genes are often transferred at an early stage of their development so that the organism develops with **desired characteristics**.

Key Words

Clone • Embryo • Enzyme • Gene • Hormone • Insulin

Insulin Production

Insulin is the **hormone** that helps to control the level of glucose in your blood.

Diabetics can't produce enough insulin and often need to inject it.

Human insulin genes can be produced by genetic engineering.

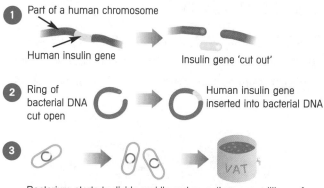

1 Part of a human chromosome

Human insulin gene

Insulin gene 'cut out'

2 Ring of bacterial DNA cut open

Human insulin gene inserted into bacterial DNA

3

VAT

Bacterium starts to divide rapidly and soon there are millions of them, each with instructions to make insulin.

The Great Genetics Debate

Scientists have made great advances in their understanding of genes:

- They have identified genes that control certain characteristics.
- They can determine whether a person's genes might increase the risk of them contracting a particular illness.
- They might soon be able to 'remove' faulty genes.

But, some people are concerned that…

- parents will want to choose the genetic make-up of their children
- unborn babies will be aborted if their genetic make-up is faulty
- insurance companies will genetically screen applicants and refuse to insure people who have an increased genetic risk of illness or disease.

Evolution

The Theory of Evolution

The **theory of evolution** states that all living things which exist today, and many more that are now **extinct, evolved** from **simple life forms** that first developed three billion years ago.

Studying the similarities and differences between species can help us to understand evolution.

Fossils are the **remains of plants or animals** from many years ago which are found in rock. They provide **evidence** of how organisms have changed over time.

Evolution by Natural Selection

Evolution is the change in a population over many generations. It may result in the formation of a new species. The members of the new species are **better adapted** to their environment.

For example, peppered moths were originally pale in colour, which meant they were **camouflaged** against the bark of silver birch trees and **predators** found it hard to see them.

But during the industrial revolution the air became polluted and silver birch trees turned black with soot. **Natural selection** led to a new variety of peppered

moth which was darker and so, better camouflaged against the darker trees.

1. Individual organisms within a species show **variation**.
2. Individuals better adapted to their environment are more likely to survive, breed successfully and produce offspring. This is **survival of the fittest**.
3. These survivors will pass on their genes to their offspring, resulting in an improved organism being **evolved** through **natural selection**.

Where new forms of a gene result from **mutation**, there may be a more rapid change in a species.

Peppered Moth

Dark Peppered Moth

Extinction of Species

The reasons for the **extinction** of species include…
- new / increased competition
- changes in the environment
- new predators
- new diseases.

The Population Explosion

The standard of living for most people has improved a lot over the past 50 years. The **human population** is now **increasing exponentially** (i.e. with accelerating speed).

This rapid increase in the human population means that…

* raw materials, including **non-renewable energy resources**, are being used up quickly
* more and more **waste** is being produced (so more landfill sites are needed)
* improper handling of waste is causing **pollution**
* there is less land available for plants and animals due to…
  – farming
  – quarrying
  – building
  – dumping waste.

Pollution

Human activities may pollute…
* **water** – with sewage, fertiliser or toxic chemicals
* **air** – with smoke and gases, such as carbon dioxide, sulfur dioxide and oxides of nitrogen which contribute to acid rain
* **land** – with toxic chemicals like **pesticides** and **herbicides** which may be washed from land into water.

Unless waste is properly handled and stored, more pollution will be caused.

Indicators of Pollution

Living organisms can be used as **indicators of pollution**, for example…
* the absence of **lichens** (a blend of algae and fungus) can indicate air pollution
* the presence or absence of invertebrate animals can indicate water pollution, e.g. freshwater shrimp survive only in unpolluted water.

Key Words

Evolve • Exponentially • Extinct • Fossil • Herbicide • Mutation • Natural selection • Non-renewable • Pesticide • Pollution • Variation

The Greenhouse Effect

Deforestation

Deforestation involves the large-scale cutting down of trees...

- for timber
- to provide land for agricultural use.

Deforestation has occurred in many tropical areas with **devastating consequences** for the environment.

Deforestation has...

- increased the amount of carbon dioxide (CO_2) released into the atmosphere (due to burning of wood, and decay of wood by microorganisms)
- reduced the rate at which carbon dioxide is removed from the atmosphere by **photosynthesis**.

Deforestation reduces **biodiversity** and results in the loss of organisms that could be of future use.

The Greenhouse Effect

Rays from the Sun reach Earth and are reflected back towards the atmosphere

CO_2 and CH_4 in the atmosphere absorb some of the energy and radiate it back to Earth

The **greenhouse effect** describes how gases in the atmosphere, such as **methane** and carbon dioxide, prevent too much heat 'escaping' from the Earth's surface into space.

The levels of these gases are slowly rising because...

- increases in the numbers of cattle and rice fields mean more methane (CH_4) is being released into the atmosphere
- the burning of chopped-down wood and industrial burning means more carbon dioxide is being released into the atmosphere.

As a result, more heat is radiated back to Earth. This is causing **global warming**.

A rise in global temperature by only a few degrees Celsius could lead to...

- substantial climate changes
- a rise in sea level.

Sustainable Development

Sustainable development means improving the quality of life on Earth without compromising future generations. It needs to be considered at…

- **local** levels
- **regional** levels
- **global** levels.

Sustainable development is concerned with three related issues:

- **economic development**
- **social development**
- **environmental protection**.

Sustainable resources are resources that can be **maintained** in the long term at a level that allows appropriate consumption / use by people.

This often requires **limiting exploitation** by using **quotas**, or by ensuring the resources are **replenished / restocked**.

Example 1

The UK has one of the largest sea fishing industries in Europe.

To ensure the industry can continue and fish stocks can be conserved, quotas are set to prevent over-fishing:

- Mesh size has been increased to prevent young fish being caught before they reach breeding age.
- Quotas of types of fish other than cod have been increased.

Example 2

Scandinavia uses a lot of pine wood to make furniture and paper, and to provide energy.

To ensure the long-term economic viability of pine-related industries, companies restock the pine forests by planting a new sapling for each tree they cut down.

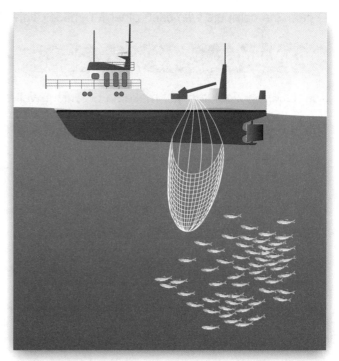

Key Words

Biodiversity • Deforestation • Global warming • Greenhouse effect • Methane • Sustainable

Unit 1b Summary

Competition

Organisms compete with each other for space / light, food and water.

Adaptations

Adaptations = Features that make an organism well-suited to its environment.

Those which are better adapted usually exist in larger numbers.

Genetics

Cell nucleus contains **chromosomes**. **DNA** makes up chromosomes. A section of a chromosome is a **gene**. Humans have 23 pairs of chromosomes.

Variation

Variation = Differences between individuals of the same species.

May be due to genetic factors or environmental factors.

Sexual reproduction ➡ Variation

Asexual reproduction ➡ No variation ➡ **Clones**

Cloning Techniques

Cloning Technique	Process	Result
Tissue culture	Cells scraped from plants and used to grow new plants.	Offspring identical to parent plant.
Embryo transplants	Cells split before they are specialised. Embryos transplanted into host mothers.	Offspring identical to each other (not to parents).
Adult cell / fusion cell cloning	DNA from donor put into empty egg cell and allowed to develop into embryo. Embryo implanted in another female.	Offspring identical to donor.

Genetic Modification

Genetic modification = Transferring genetic material from one organism to another. Genes cut out using **enzymes**.

Insulin can be made by genetic engineering:

Human insulin gene cut out ➡ Bacterial DNA cut open ➡ Human insulin gene inserted ➡ Bacterium divides to produce millions of insulin-producing bacteria

Evolution

Theory of evolution states that all living things developed from simple life forms.

Fossils = Remains of plants / animals found in rock. Fossils provide evidence of evolution.

Evolution = Change in population over many generations. May result in new species, which is better adapted to its environment.

Individuals in a population show variation → Survival of the fittest → Survivors pass on genes to offspring → Evolution of better adapted organism

Species may become **extinct** due to...
- new / increased competition
- changes in environment
- new predators
- new diseases.

Pollution

Human population is increasing exponentially, leading to...
- raw materials being used up
- more waste produced
- less land (due to farming, quarrying, building, dumping waste)
- **pollution** (of water, air and land).

Deforestation

Deforestation = Large-scale cutting down of trees for timber and agriculture.

Deforestation → increases amount of CO_2 released into atmosphere.

→ reduces rate at which CO_2 is removed from atmosphere.

→ reduces **biodiversity**.

Greenhouse Effect

Greenhouse effect describes how gases in the atmosphere prevent too much heat escaping from Earth into space.

Levels of greenhouse gases (e.g. methane, CO_2) rising → More heat radiated back to Earth → Global warming

Sustainable development = Improving quality of life without compromising future generations.

Unit 1b Practice Questions

1 Match each organism A, B, C and D with the statement 1 to 4 that best describes its adaptation.

A cacti .... **B** desert fox ....

C polar bear .... **D** blue dart frog ....

1 Large ears to allow the blood to cool down. **2** Bright colour and poisonous skin.

3 Has spikes for protection. **4** Thick layer of fat for insulation.

2 Name two things that organisms compete with each other for.

i) .... **ii)** ....

3 Match the words A, B, C and D with the spaces numbered 1 to 4 in the sentences below.

A genes .... **B** chromosomes ....

C DNA .... **D** inherited ....

The nucleus of a cell contains ...**1**... which are made up of a substance called ...**2**... Chromosomes are split up into sections called ...**3**... which control the characteristics of an organism. During reproduction different genes are ...**4**... from each parent.

4 This question has three parts. For each part, put a tick next to the correct answer.

a) Which of these processes doesn't involve genetic engineering?

   i) Embryo transplants

   ii) Tissue culture

   iii) Fusion cell cloning

   iv) Sexual reproduction

b) When an organism is cloned using the technique of embryo transplants…

   i) the offspring are identical to the parents.

   ii) the offspring are identical to each other and to the parents.

   iii) the offspring are identical to each other but not the parents.

   iv) the offspring are all different.

c) Cloning occurs naturally. Which process doesn't produce clones?

   i) Taking cuttings

   ii) Asexual reproduction

   iii) Sexual reprocation

   iv) Spider plant stolons

5 What is meant by the term 'variation'?

....

6 Describe the process of cloning a plant.

....

....

7 Suggest three reasons for altering an organism's genetic make-up.

i) ....

ii) ....

iii) ....

8 a) What is the theory of evolution?

....

b) What are fossils?

....

9 Match the words A, B, C and D with the spaces numbered 1 to 4 in the sentences below.

A environment .... **B** predators ....

C diseases .... **D** competition ....

Extinction happens when there is increased __1__ between organisms or if the __2__ changes too quickly. The introduction of new __3__ or new __4__ can also cause organisms to become extinct.

10 Match the words A, B, C and D with the descriptions numbered 1 to 4 below.

A deforestation .... **B** methane ....

C burning fossil fuels .... **D** carbon dioxide ....

1 Produced by herds of cattle and rice fields.
2 Acts like an insulating blanket trapping heat energy.
3 Reduces biodiversity.
4 Increases the amount of carbon dioxide released into the atmosphere.

11 List the three issues that sustainable development is concerned with.

i) ....

ii) ....

iii) ....

Cells

Cells

All **living things** are made up of **cells**. The **structures** of different types of cells are related to their **functions**.

Cells may be **specialised** in order to carry out a particular job, as the table shows:

Chemical reactions inside cells are controlled by **enzymes**.

Enzymes are found in **cytoplasm** and **mitochondria**.

Root hair cells	Ovum (egg cell)	Xylem	White blood cells	Sperm cells	Palisade cells	Red blood cells	Nerve cells
Tiny hair-like structures which increase surface area of the cell.	Large cell which can carry massive food reserves for the developing embryo.	Long, thin, hollow cells used to transport water through the stem and root.	Can change shape in order to engulf and destroy invading microorganisms.	Has a tail which allows it to move.	Packed with chloroplasts for photosynthesis.	No nucleus, so packed full of haemoglobin.	Long, slender axons which can carry nerve impulses.

Animal Cells

Human cells, most animal cells, and plant cells have the following parts:

- **Nucleus** – controls the activities of the cell.
- **Cytoplasm** – where most chemical reactions take place.
- A **cell membrane** – controls the passage of substances in and out of the cell.
- **Mitochondria** – where most energy is released in **respiration**.
- **Ribosomes** – where protein synthesis occurs.

A Human Cheek Cell

Cell membrane

Cytoplasm (may contain mitochondria)

Nucleus

Ribosomes

Plant Cells

Plant cells also have the following parts:

- A **cell wall** – used to strengthen the cell.
- **Chloroplasts** – absorb light energy to make food.
- A **permanent** vacuole – filled with cell sap.

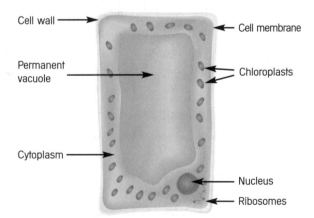

A Palisade Cell from a Leaf

Cell wall

Cell membrane

Permanent vacuole

Chloroplasts

Cytoplasm

Nucleus

Ribosomes

Key Words

Cell • Chloroplast • Cytoplasm • Diffusion • Enzyme • Mitochondria • Nucleus • Osmosis • Respiration • Ribosomes • Specialised • Vacuole

The Movement of Substances

Cells have to constantly…

- **replace** substances which are used up, e.g. food and oxygen
- **remove** other substances, e.g. carbon dioxide and waste products.

These substances can pass into and out of cells by…

- **diffusion**
- **osmosis**.

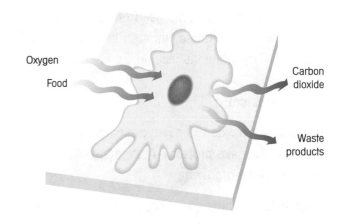

Oxygen
Food
Carbon dioxide
Waste products

Diffusion

Diffusion is the spreading of the **particles of a gas**, or any substance in solution, which results in a **net movement** from a region where they are at a **higher concentration**, to a region where they are at a **lower concentration**.

Oxygen required for respiration passes through cell membranes by diffusion. The **greater the difference** in concentration, the **faster the rate** of diffusion.

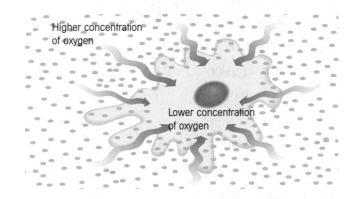

Higher concentration of oxygen

Lower concentration of oxygen

Osmosis

Osmosis is the **movement of water** from a **dilute solution** to a **more concentrated solution** through a partially permeable membrane.

The membrane allows the passage of water molecules, but not solute molecules because they are too large.

Osmosis will gradually **dilute** the more concentrated solution. For example, at root hair cells water moves from the soil into the cell by osmosis, along a concentration gradient.

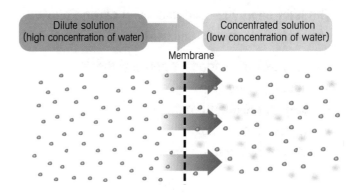

Dilute solution (high concentration of water)
Concentrated solution (low concentration of water)
Membrane

Less concentrated solution (dilute)

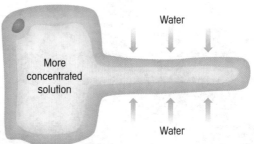

Water

More concentrated solution

Water

Photosynthesis

Plant Mineral Requirements

For healthy growth, plants need **mineral ions** which they absorb from the soil through their roots. The necessary mineral ions include…

- **nitrates**
- **magnesium**.

Nitrates are needed to make amino acids which are used to make **proteins**. A shortage of nitrates leads to stunted growth.

Magnesium is needed for **chlorophyll production**. A shortage leads to yellow leaves.

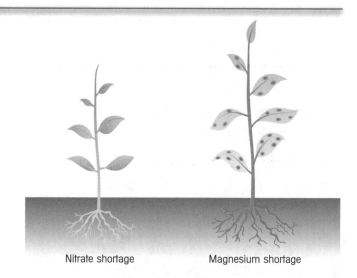

Nitrate shortage   Magnesium shortage

Photosynthesis

Green plants don't absorb food from the soil. They make their own using sunlight. This process is called **photosynthesis**.

Photosynthesis occurs in the cells of **green plants** which are exposed to light.

During photosynthesis, light energy is absorbed by green **chlorophyll** which is found in chloroplasts in some plant cells.

The **light energy** is used to convert **carbon dioxide** and **water** into **sugar** (glucose).

Four things are needed for photosynthesis:
- **light** from the Sun
- **carbon dioxide** from the air
- **water** from the soil
- **chlorophyll** in the leaves.

Two things are produced:
- **sugar** (glucose)
- **oxygen** – released into the atmosphere as a by-product.

The word equation for photosynthesis is…

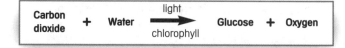

The glucose produced in photosynthesis can be…
- used by the plant to provide **energy** for **respiration**
- converted into **insoluble starch** which is stored in the stem, leaves or roots.

Factors Affecting Photosynthesis

Factors Affecting Photosynthesis

There are several factors that may at any time **limit the rate** of photosynthesis:

- **temperature**
- **carbon dioxide** (CO_2) concentration
- **light intensity**.

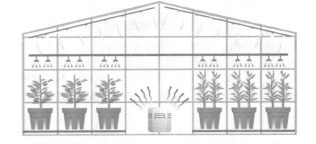

Temperature

1. As the temperature increases, so does the rate of photosynthesis. Temperature is limiting the rate of photosynthesis.
2. As the temperature approaches 45°C, the rate of photosynthesis drops to zero. The enzymes controlling photosynthesis have been destroyed.

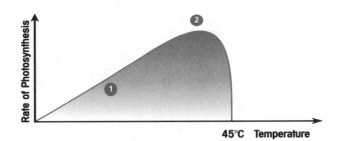

Carbon Dioxide Concentration

1. As the rate of carbon dioxide concentration increases, so does the rate of photosynthesis. CO_2 is limiting the rate of photosynthesis.
2. After reaching a certain point, an increase in carbon dioxide has no further effect. Carbon dioxide is no longer the limiting factor – it must be either light or temperature.

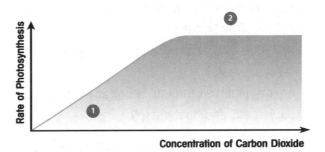

Light Intensity

1. As the light intensity increases, so does the rate of photosynthesis. Light intensity is limiting the rate of photosynthesis.
2. After reaching a certain point, an increase in light intensity has no further effect. Light intensity is no longer the limiting factor – it must be carbon dioxide or temperature.

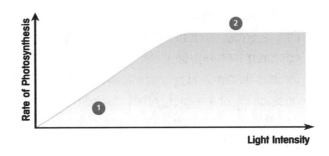

Artificial Controls

Greenhouses can be used to **control the rate** of photosynthesis. This can result in plants...

- growing more quickly
- becoming bigger and stronger.

Key Words

Chlorophyll • Mineral ions • Nitrates • Photosynthesis

Food Chains and Biomass

Food Chains

Radiation from the Sun is the source of **energy** for all **communities** of living organisms.

In green plants, photosynthesis captures a **small fraction** of the solar energy which reaches them.

This energy is stored in the substances which make up the cells of the plant. It can then be passed onto organisms which eat the plant. This transfer of energy can be represented by a **food chain**.

Grass   Rabbit   Stoat   Fox

Pyramid of Biomass

Biomass is the mass of living materials. In a food chain, the biomass at each stage is **less** than it was at the previous stage. The biomass at each stage can be drawn to scale and shown as a **pyramid of biomass**.

Biomass and energy are **lost** at every stage of a food chain. This is due to…

* materials and energy being lost in an organism's faeces (waste)
* energy released through **respiration** being lost in movement and heat energy.

Mammals and birds, in particular, lose a lot of biomass. They must keep their body temperature constant, and higher than their surroundings.

In the biomass pyramid:

* Only a small amount of the Sun's energy is captured by the producers.
* Rabbits respire and produce waste products. They pass on $\frac{1}{10}$ of the energy they get from the grass.
* Stoats respire and produce waste products. They pass on $\frac{1}{10}$ of the energy they get from the rabbits.
* The fox gets the last bit of energy and biomass.

Since the loss of energy and biomass is due mainly to **heat loss**, **waste** and **movement**, the efficiency of **food production** can be improved by…

* reducing the number of stages in a food chain
* limiting an animal's movement
* controlling an animal's temperature.

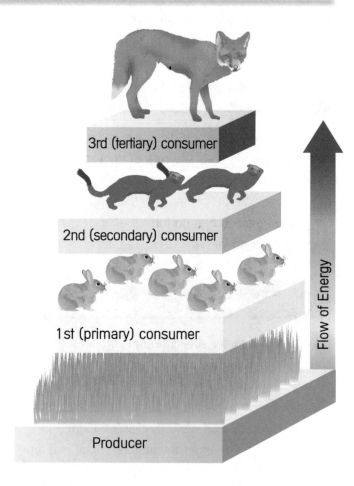

3rd (tertiary) consumer

2nd (secondary) consumer

1st (primary) consumer

Producer

Flow of Energy

Key Words

Biomass • Carbon cycle • Community • Decay • Detritus • Environment • Food chain • Photosynthesis • Radiation • Respiration

Recycling the Materials of Life

Living things (organisms) **remove materials** from the **environment** for growth and other processes.

When the organisms die or excrete waste, the materials are **returned** to the environment.

Microorganisms break down waste and dead bodies. This **decay process** releases substances needed by plants for growth.

Microorganisms digest materials faster in conditions that are **warm**, **moist** and have plenty of **oxygen**.

Microorganisms are used in…
- sewage works to break down **human waste**
- compost heaps to break down **plant material**.

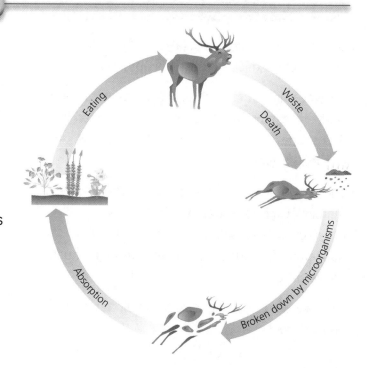

The Carbon Cycle

In a stable community, two processes are **balanced**:
- the **removal** of materials from the environment
- the **return** of materials to the environment.

This constant recycling of carbon material is called the **carbon cycle**.

1. Carbon dioxide, CO_2, is **removed** from the atmosphere by green plants for **photosynthesis**. Some CO_2 is **returned** to the atmosphere by plants during **respiration**.

2. The carbon obtained by photosynthesis is used by **plants** to make carbohydrates, fats and proteins. The plants are eaten by **animals**. Some of the carbon becomes fats and proteins in the animals.

3. Animals **respire**, releasing CO_2 into the **atmosphere**.

4. Plants and animals **die**. Other animals and microorganisms feed on their bodies, causing them to **break down**.

5. **Detritus** feeders and microorganisms respire, releasing CO_2 into the atmosphere.

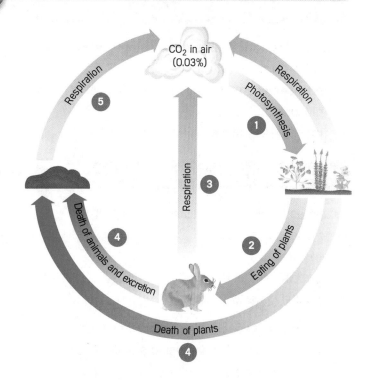

Enzymes

Enzymes

Enzymes are **biological catalysts**. They increase the rate of chemical reactions in an organism.

Enzymes are **protein** molecules, made up of long chains of **amino acids**. The chains are folded in a 3-D shape. This shape allows other molecules to fit into the enzyme.

The **shape** of the enzyme is **vital** for the enzyme's **function**. **High temperatures destroy** most enzymes' special shape. This is why it's dangerous for your body temperature to go much above 37°C.

Different enzymes work best at certain temperatures and pH levels.

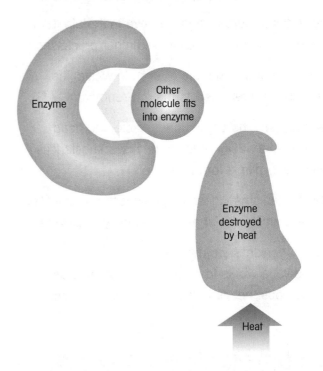

Aerobic Respiration

Aerobic respiration takes place **using oxygen**. This type of respiration…
- releases energy through the breakdown of glucose molecules
- mostly takes place inside mitochondria.

The written equation for aerobic respiration is:

$$\text{Glucose} + \text{Oxygen} \xrightarrow[\text{by enzymes}]{\text{catalysed}} \text{Carbon dioxide} + \text{Water} \; (+\text{Energy})$$

Enzymes Inside Living Cells

Enzymes in living cells catalyse (speed up) processes such as…
- respiration
- protein synthesis
- photosynthesis.

The energy released during respiration is used to…
- build larger **molecules** using smaller ones
- enable **muscles** to contract (in animals)
- **maintain a constant temperature** (in mammals and birds)
- **make proteins** in plants from amino acids (from sugars and nitrates).

N.B. They all begin with M, so remember the four 'M's.

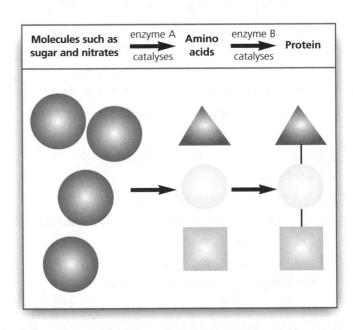

Enzymes Outside Living Cells

Some enzymes work **outside** the body cells.
Digestive enzymes are produced by **specialised** cells in glands in the digestive system.

The enzymes…
1. Pass out of the cells into the digestive system.
2. Come into contact with food molecules.
3. Catalyse the breakdown of large food molecules into smaller molecules.

Three enzymes (**protease**, **lipase**, and **amylase**) are produced in four separate regions of the digestive system (**salivary glands**, **stomach**, **pancreas**, and **small intestine**).

The enzymes **digest** proteins, fats and carbohydrates to produce **smaller molecules** which can be **absorbed**.

Amylase…
- is produced in the salivary glands, pancreas and small intestine
- digests starch
- produces sugars.

Protease…
- is produced in the stomach, pancreas and small intestine
- digests proteins
- produces amino acids.

Lipase…
- is produced in the pancreas and small intestine
- digests lipids (fats and oils)
- produces fatty acids and glycerol.

Salivary Glands

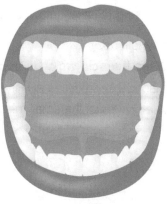

Stomach
Also produces hydrochloric acid in which these enzymes work best.

Pancreas
Produces enzymes that are released into the small intestine.

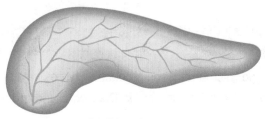

Small Intestine

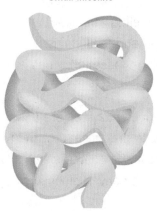

Key Words

Aerobic respiration • Amylase • Catalyst • Enzyme • Lipase • Protease • Specialised

Enzymes

Bile

Bile is **produced** in your **liver**. It is **stored** in your **gall bladder** before being released into your **small intestine**.

Bile has two functions:

- **It neutralises the acid** which is added to food in your stomach. This produces alkaline conditions in which the **enzymes** of the small intestine work best.

- **It emulsifies fats** (breaks down large drops of fat into small droplets to increase their surface area). This enables the **lipase** enzymes to work much faster.

Drops of fat → Bile → Droplets of fat

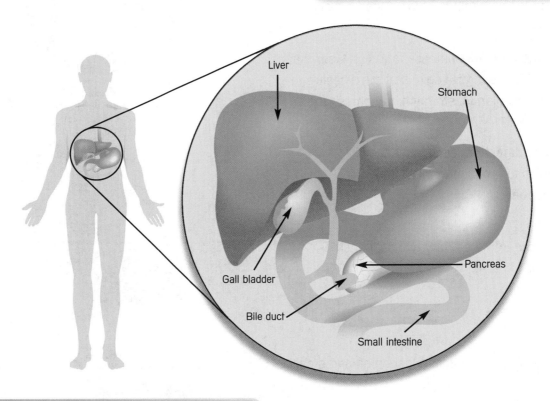

Liver

Stomach

Pancreas

Gall bladder

Bile duct

Small intestine

Use of Enzymes

Some microorganisms produce enzymes which can be used to our benefit both **in the home** and **in industry**.

In the home, biological detergents may contain…

- **protein-digesting** (protease) enzymes, to break down blood and food stains
- **fat-digesting** (lipase) enzymes, to break down oil and grease stains.

In industry, various enzymes are used, such as…

- **proteases**, to pre-digest protein in baby foods
- **carbohydrases**, to convert starch into sugar syrup
- **isomerase**, to convert glucose syrup into fructose syrup.

Controlling Body Conditions

In order to function properly, your body has to control levels of...

- blood sugar
- water content
- **ion** content
- temperature.

Blood Glucose Concentration

Blood glucose concentration is monitored and controlled by the **pancreas** which secretes the hormone **insulin**.

The level of insulin in the pancreas affects what happens in the **liver**. The pancreas...

- continually monitors the body's blood sugar levels
- adjusts the amount of insulin released in order to keep the body's blood sugar levels as close to normal as possible.

If the pancreas doesn't produce enough insulin, a person's blood glucose concentration may rise to a fatally high level.

This is a condition called **diabetes**, which can be controlled by...

- careful attention to diet
- injecting insulin into the blood.

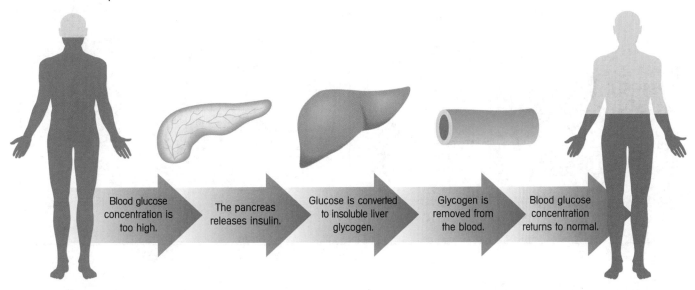

| Blood glucose concentration is too high. | The pancreas releases insulin. | Glucose is converted to insoluble liver glycogen. | Glycogen is removed from the blood. | Blood glucose concentration returns to normal. |

Water and Ion Content

If the **water** or **ion content** of the body is wrong, too much water may move into or out of the cells by **osmosis**. This can cause damage to the cells.

Your body gains water and ions through food and drink.

Key Words

Bile • Diabetes • Enzyme • Insulin • Ion • Osmosis

Body Temperature

Body Temperature

Your body temperature is **monitored** and **controlled** by the **thermoregulatory centre** in the brain. This centre has receptors that are sensitive to the temperature of blood flowing through them.

Temperature receptors in the skin also provide information about skin temperature. Sweat helps to cool your body when it is **hot**. More water is **lost** from the body, so more water has to be **taken** in as food or drink to balance this.

Your body temperature should be kept at around 37°C.

HT Hot Conditions:
If your core body temperature becomes too **high**…
- blood vessels in skin **dilate** (become wide), increasing heat loss
- sweat glands release **sweat** which evaporates, causing cooling.

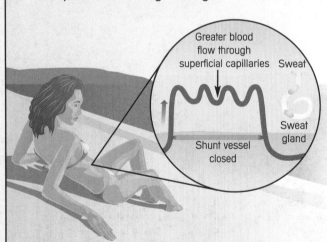

Greater blood flow through superficial capillaries

Sweat

Sweat gland

Shunt vessel closed

HT Cold Conditions:
If your core body temperature becomes too **low**…
- blood vessels in skin **constrict** (become narrower), reducing heat loss
- muscles start to **'shiver'** causing heat energy to be released by respiration in cells.

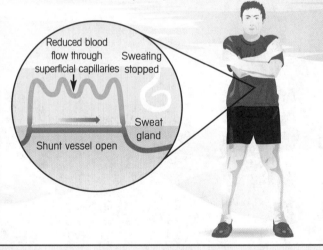

Reduced blood flow through superficial capillaries

Sweating stopped

Sweat gland

Shunt vessel open

Removing Waste Products

You need to **remove waste products** from your body to keep your internal environment relatively **constant**.

Two of these waste products are…
- **carbon dioxide** – produced by **respiration** and removed through the lungs during exhalation
- **urea** – produced by the liver when it breaks down amino acids, removed by the kidneys, and temporarily stored in the bladder.

Chromosomes and Gametes

Human Body Cells

Body cells contain **46 chromosomes** arranged as **23 pairs**.

Gametes are sex cells, i.e. female eggs and male sperm. Gametes have **23 chromosomes** (one from each pair).

The fusion of two gamete cells produces a **zygote**. A zygote has 46 chromosomes in total (23 pairs).

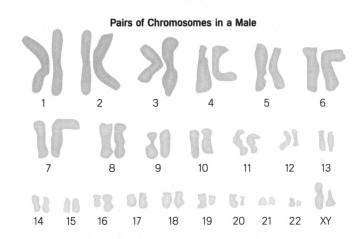

Pairs of Chromosomes in a Male

1 2 3 4 5 6
7 8 9 10 11 12 13
14 15 16 17 18 19 20 21 22 XY

Inheritance of the Sex Chromosome

Of the 23 pairs of chromosomes in the human body, 1 pair contains the **sex chromosomes**.

- **In females**, these chromosomes are identical and are called the X chromosomes.
- **In males**, one chromosome (Y) is much shorter than the other (X).

Female Sex Chromosomes	Male Sex Chromosomes
X X	X Y

Like all pairs of chromosomes, offspring inherit…
- one sex chromosome from the mother
- one sex chromosome from the father.

So, the sex of an individual is ultimately decided by whether the ovum is fertilised by an **X-carrying sperm** or a **Y-carrying sperm**.

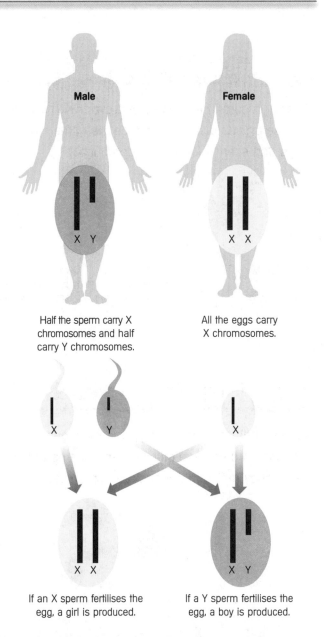

Male

Female

Half the sperm carry X chromosomes and half carry Y chromosomes.

All the eggs carry X chromosomes.

X   Y   X

If an X sperm fertilises the egg, a girl is produced.

If a Y sperm fertilises the egg, a boy is produced.

Key Words

Chromosome • Gamete • Respiration • Thermoregulation • Urea • Zygote

Cell Division

Mitosis

Mitosis is the **division** of body cells to produce new cells. Mitosis occurs…

- for growth
- for repair
- in asexual reproduction.

Before the cell divides, a copy of each chromosome is made. The **new cell** contains exactly the **same genetic information** as the **parent** cell, i.e. the same number of chromosomes and the same genes.

Parent cell with two pairs of chromosomes.

Each chromosome replicates itself.

The copies are pulled apart. Cell now divides for the only time.

Two 'daughter' cells are formed.

HT Meiosis

Meiosis occurs in the **testes** and **ovaries**. The cells in these organs divide to produce the gametes (eggs and sperm) for sexual reproduction.

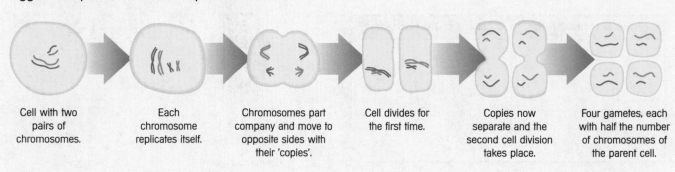

Cell with two pairs of chromosomes.

Each chromosome replicates itself.

Chromosomes part company and move to opposite sides with their 'copies'.

Cell divides for the first time.

Copies now separate and the second cell division takes place.

Four gametes, each with half the number of chromosomes of the parent cell.

Fertilisation

During fertilisation, the female and male gametes join. One chromosome comes from each parent. A **single body cell** with **new pairs** of chromosomes is

formed. The cell then divides repeatedly by **mitosis** to form a new individual, giving rise to **variation**.

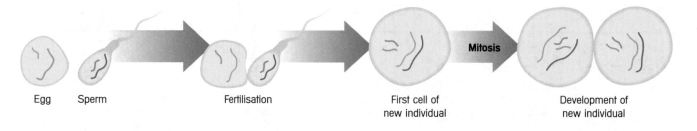

Egg   Sperm

Fertilisation

First cell of new individual

Mitosis

Development of new individual

Alleles

Some **genes** have different forms or variations, called **alleles**. For example…

- the gene that controls tongue-rolling ability has two alleles – you either can or you can't
- the gene that controls eye colour has two alleles – blue or brown.

In a pair of chromosomes, the alleles for a gene can be the **same** or **different**. If they are different, then one allele will be **dominant** and one allele will be **recessive**.

A dominant allele **will control** the characteristics of the gene. Dominant alleles express themselves even if present on only one chromosome. So, in the example of brown eyes, an individual can be…

- **homozygous dominant** (BB)
- **heterozygous** (Bb).

A recessive allele will **only** control the characteristics of the gene if it is present on **both chromosomes in a pair** (i.e. no dominant allele is present). So, in the example of blue eyes, an individual can only be **homozygous recessive** (bb).

Example: Dominant and Recessive Alleles
The diagram shows three pairs of genes from the middle of a pair of chromosomes. The genes code for…

- tongue-rolling ability
- eye colour
- type of earlobe (i.e. attached or unattached).

Key Words

Allele • Fertilisation • Gene • Meiosis • Mitosis

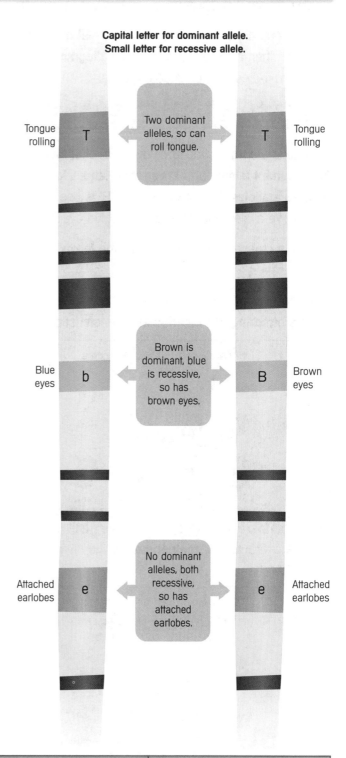

Capital letter for dominant allele.
Small letter for recessive allele.

Tongue rolling — T — Two dominant alleles, so can roll tongue. — T — Tongue rolling

Blue eyes — b — Brown is dominant, blue is recessive, so has brown eyes. — B — Brown eyes

Attached earlobes — e — No dominant alleles, both recessive, so has attached earlobes. — e — Attached earlobes

	Homozygous Dominant	Heterozygous	Homozygous Recessive
Tongue rolling	TT (can roll)	Tt (can roll)	tt (can't roll)
Eye colour	BB (brown)	Bb (brown)	bb (blue)
Earlobes	EE (free earlobes)	Ee (free earlobes)	ee (attached earlobes)

Genetic Diagrams

Monohybrid Inheritance

When a characteristic is determined by **just one pair of alleles** then a simple **genetic cross diagram** can be drawn to show inheritance of genes.

This type of inheritance is referred to as **monohybrid inheritance**.

Inheritance of Eye Colour

In genetic diagrams you should use **capital letters for dominant alleles** and **lower case letters for recessive alleles**. So, for eye colour, 'B' is used for brown eye alleles and 'b' is used for blue eye alleles.

From the crosses on the diagrams, the following can be seen:

1 If one parent has two dominant genes then **all the offspring** will inherit that characteristic.

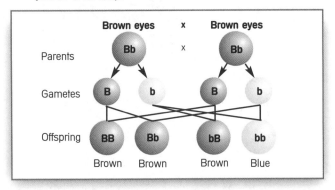

2 If both parents have one recessive gene then this characteristic **may** appear in the offspring (25% chance).

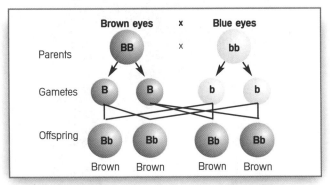

3 If one parent has one recessive gene and the other has two recessive genes, then there is a 50% chance of that characteristic appearing in the offspring.

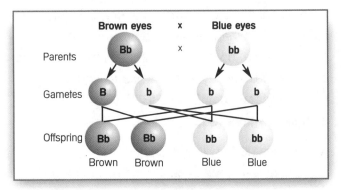

But remember, these are **only probabilities**. In practice, all that matters is which egg is fertilised by which sperm. This process is completely **random**.

HT When you construct genetic diagrams remember to...
- clearly **identify** the alleles of the parents
- place each of these alleles in a **separate gamete**
- **join** each gamete with the two gametes from the other parent.

Differentiation of Cells

Differentiation is when cells develop a **specialised** structure to carry out a specific function.

- Most plant cells can differentiate throughout life.
- Animal cells **can only** differentiate at an early stage.
- Mature animal cells divide for repair / replacement.

Chromosomes, DNA and Genes

Stem Cells

Stem cells are cells in human **embryos** and adult **bone marrow** which have **not yet differentiated**.

They can differentiate into **many different types** of cells, e.g. nerve cells. Treatment using these cells may help treat people with conditions like **paralysis**.

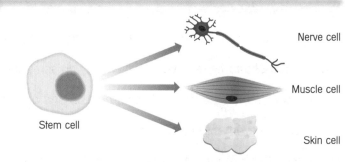

Nerve cell

Muscle cell

Stem cell

Skin cell

Chromosomes, DNA and Genes

Chromosomes are made up of long, coiled molecules of **DNA** (**deoxyribonucleic acid**).

A DNA molecule consists of **two strands** which are coiled to form a **double helix**.

Each person has **unique** DNA (apart from identical twins). So, DNA can be used for **identification** (DNA fingerprinting).

A **gene** is a small section of DNA. Genes code for a particular inherited characteristic, e.g. blue eyes.

HT Genes code for a particular characteristic by providing a code for a combination of amino acids which make up a protein.

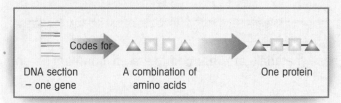

Codes for

DNA section
– one gene

A combination of amino acids

One protein

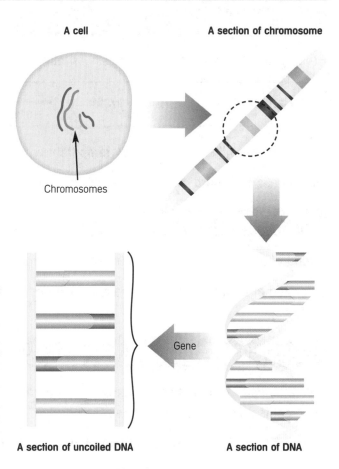

A cell

A section of chromosome

Chromosomes

Gene

A section of uncoiled DNA

A section of DNA

Genetic Disorders

Embryos can be screened for the genes that cause genetic disorders, such as…
*   Huntington's disease
*   Cystic fibrosis.

Huntington's disease is a disorder of the nervous system caused by a **dominant allele**. It can be passed on even if only **one parent** has the disorder.

Cystic fibrosis is a disorder of cell membranes. It is caused by a **recessive allele**. So, it must be inherited from **both parents**. The parents can be **carriers** without having the disorder themselves.

Key Words

Allele • Differentiation • DNA • Gene • Specialised • Stem cell

Unit 2 Summary

Cells

The structure of cells is related to their **function**. Cells may be **specialised** to carry out a particular job.

Animal and plants cells have a nucleus, cytoplasm, cell membrane, mitochondria and ribosomes.

Plant cells also have a cell wall, chloroplasts and a permanent vacuole.

Diffusion = Movement of **particles** from high to low concentration.

Osmosis = Movement of **water** from high to low concentration.

Plant Growth

Green plants **make** their own **food** using **sunlight**. This process is called **photosynthesis**.

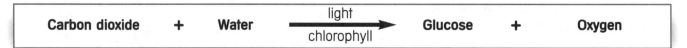

Carbon dioxide	+	Water	light / chlorophyll →	Glucose	+	Oxygen

Rate of photosynthesis is limited by **temperature**, **carbon dioxide concentration** and **light intensity**.

Plants need to **absorb mineral ions** (nitrates and magnesium) for healthy growth.

Food Chains and Biomass

A **food chain** represents the **transfer of energy** from organism to organism.

Pyramids of biomass show the mass of **living materials** at each stage of the food chain. Biomass is **lost** at each stage.

Recycling Materials

Living things **remove materials** from the environment. These materials are **returned** to the environment when the organisms die or excrete waste.

The **carbon cycle** is the constant **recycling** of carbon material.

Aerobic Respiration

Aerobic respiration takes place **using oxygen**.

Glucose	+	Oxygen	catalysed by enzymes →	Carbon dioxide	+	Water	(+ Energy)

Enzymes

Enzymes are **biological catalysts**. **Inside** body cells, enzymes **speed up** reactions such as respiration, protein synthesis and photosynthesis.

Outside body cells, digestive enzymes are produced in the digestive system.

Bile, produced in the liver, neutralises acid and emulsifies fats.

Enzyme	Where Produced
Amylase	• Salivary glands • Pancreas • Small intestine
Protease	• Stomach • Small intestine
Lipase	• Pancreas • Small intestine

Controlling Body Conditions

The human body has to control the levels of blood sugar, water, ion content and temperature.

Diabetes = A condition in which the pancreas isn't able to control the levels of insulin in the body.

Body temperature is monitored and controlled by the **thermoregulatory** centre in the brain.

HT

Body Temperature	Body Reaction
Too high	• Blood vessels dilate • Sweat glands release sweat
Too low	• Blood vessels constrict • Muscles start to shiver causing heat energy to be released.

Human Body Cells

Human body cells ➡ **46 chromosomes** ➡ 23 pairs

Gametes ➡ sex cells ➡ 23 chromosomes

Two gametes join to produce a **zygote** with 46 chromosomes.

DNA ➡ Gene ➡ Chromosome

Alleles = Variations of genes.

Dominant alleles – control gene characteristics if present **once**. **Recessive** alleles – only control gene characteristics if present **twice**.

Differentiation – cells carry out a specific function. Stem cells have not yet differentiated.

Mitosis

Mitosis = **Division** of body cells to produce new cells. The new cells contain **identical genetic** information to that of the parent cell.

After **fertilisation**, a zygote divides repeatedly by mitosis to form a new individual.

HT **Meiosis** = Division of cells in the testes and ovaries to produce gametes. The gametes will have half the number of chromosomes as the parent cell.

1 The diagram shows a human cheek cell.

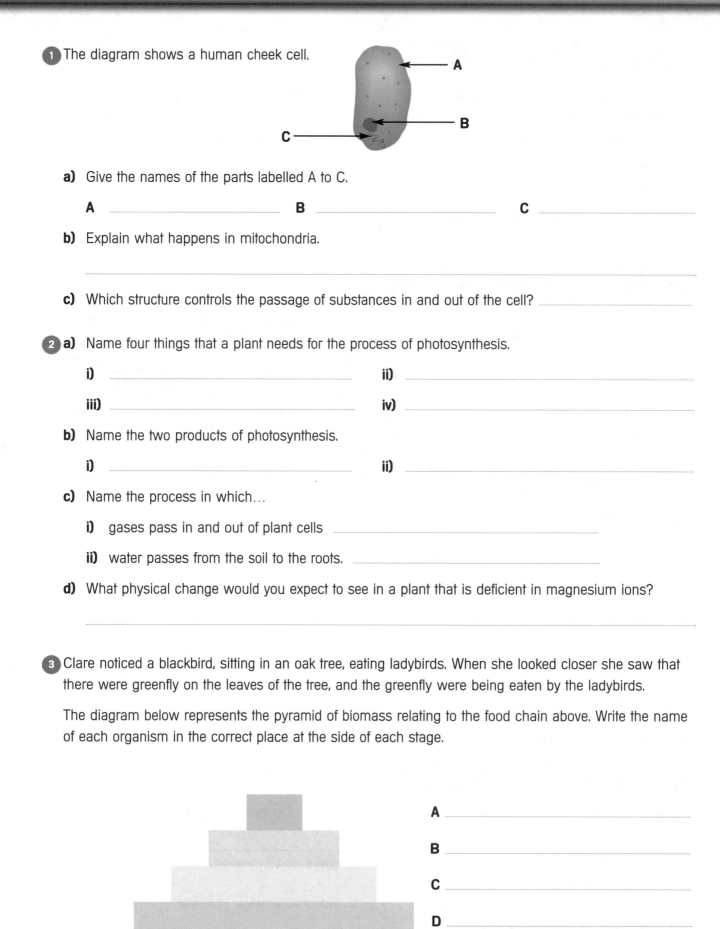

a) Give the names of the parts labelled A to C.

A .... B .... C ....

b) Explain what happens in mitochondria.

....

c) Which structure controls the passage of substances in and out of the cell? ....

2 a) Name four things that a plant needs for the process of photosynthesis.

i) .... **ii)** ....

iii) .... **iv)** ....

b) Name the two products of photosynthesis.

i) .... **ii)** ....

c) Name the process in which…

i) gases pass in and out of plant cells ....

ii) water passes from the soil to the roots. ....

d) What physical change would you expect to see in a plant that is deficient in magnesium ions?

....

3 Clare noticed a blackbird, sitting in an oak tree, eating ladybirds. When she looked closer she saw that there were greenfly on the leaves of the tree, and the greenfly were being eaten by the ladybirds.

The diagram below represents the pyramid of biomass relating to the food chain above. Write the name of each organism in the correct place at the side of each stage.

A ....

B ....

C ....

D ....

4 This question has two parts. For each part, tick the correct answer.

   a) Which organ in the human body monitors blood sugar concentration?

     i) The liver ◯   **ii)** The pancreas ◯

     iii) The kidneys ◯   **iv)** The gall bladder ◯

   b) If the body is unable to produce enough insulin naturally, what could happen to blood glucose concentration?

     i) It could rise a little ◯   **ii)** It could drop ◯

     iii) It could rise to a fatally high level ◯   **iv)** It could stay the same ◯

5 Biological catalysts are enzymes which speed up chemical reactions in an organism.

   a) Give two processes inside living cells that enzymes could speed up.

     i) ....   **ii)** ....

   b) Name two factors that can change the effectiveness of enzymes.

     i) ....   **ii)** ....

   c) Describe the functions of bile.

     ....

6 There are several parts to this question. For each part, tick the correct answer.

   a) Name the large molecule which forms the basis of chromosomes.

     i) Gene ◯   **ii)** Alleles ◯

     iii) DNA ◯   **iv)** Gametes ◯

   b) How many chromosomes are there in a normal human sperm?

     i) 22 ◯   **ii)** 23 ◯

     iii) 46 ◯   **iv)** 50 ◯

   HT c) Nick is homozygous dominant for brown eyes and Claire is homozygous recessive for blue eyes. Nick and Claire have a child. What is the percentage chance of their child having brown eyes?

     i) 25% ◯   **ii)** 50% ◯

     iii) 75% ◯   **iv)** 100% ◯

Exchanging Materials

Osmosis and Diffusion

Water and **dissolved substances** automatically move along a **concentration gradient**.

They move **from high** concentrations **to low** concentrations.

They move by **osmosis** and **diffusion**.

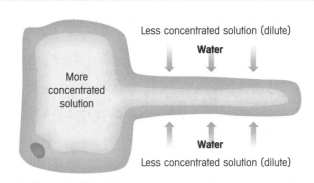

(HT) Active Transport

Substances are sometimes absorbed **against** a concentration gradient. But this means using **energy from respiration**. This is known as **active transport**. Plants absorb ions from very dilute solutions by active transport.

Active transport takes place in the **opposite direction** to normal diffusion.

Sugar and **ions**, which can pass through cell membranes, can also be moved by active transport.

In humans, sugar can be absorbed from the **intestine** and from the **kidney tubules** by active transport.

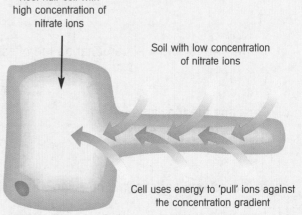

A Cell Absorbing Ions by Active Transport

Root hair cell with high concentration of nitrate ions

Soil with low concentration of nitrate ions

Cell uses energy to 'pull' ions against the concentration gradient

Exchanging Materials in Humans

Humans have organ systems which are **specialised** to help the exchange of materials. For example...
- the villi in your small intestine
- the alveoli in your lungs (part of the breathing system).

Villi in the Small Intestine

Villi line the walls of your small intestine. They have a massive **surface area** and an extensive network of **capillaries**.

This network absorbs the products of digestion by...
- diffusion
- active transport.

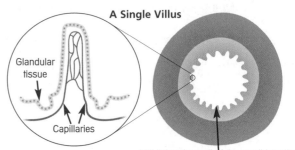

A Single Villus

Glandular tissue

Capillaries

Villi lining the wall of the small intestine

The Breathing System

The **breathing system** involves the **heart** and the **lungs**. It takes air into and out of your body.

Your **ribcage** protects your heart and lungs (i.e. the contents of your thorax).

Your thorax is divided from your abdomen by a muscular sheet called the **diaphragm**.

Air that you breathe in reaches the lungs through the **trachea** (windpipe).

The trachea has rings of cartilage to prevent it from collapsing. It divides into two tubes called the bronchi, which divide again to form **bronchioles**.

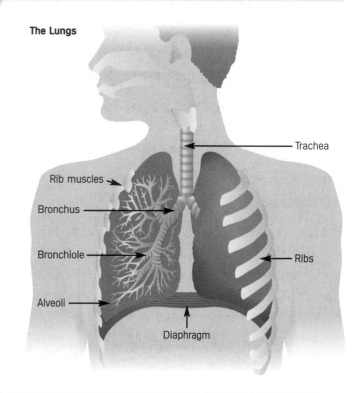

The Lungs

- Trachea
- Rib muscles
- Bronchus
- Bronchiole
- Alveoli
- Diaphragm
- Ribs

Alveoli in the Lungs

The bronchioles continue to divide until they end in air sacs called **alveoli** (there are millions of these). The alveoli are very close to the blood capillaries.

The arrangement of alveoli and capillaries in your lungs makes them efficient at exchanging **oxygen** and **carbon dioxide**. This is because they have…
- a large, moist surface area
- an excellent blood supply.

Carbon dioxide diffuses from your **blood** into your **alveoli**.

Oxygen diffuses from your **alveoli** into your **blood**.

So, your blood swaps carbon dioxide for oxygen to become **oxygenated**.

Key Words

Active transport • Alveoli • Concentration gradient • Diffusion • Osmosis • Specialised • Surface area • Villi

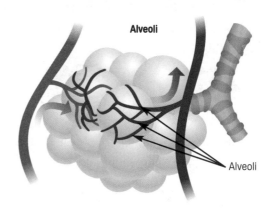

Alveoli

- Alveoli

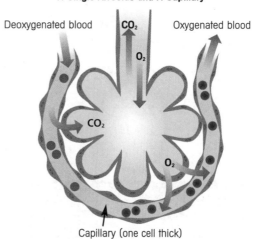

A Single Alveolus and A Capillary

Deoxygenated blood CO_2 Oxygenated blood

O_2

CO_2

O_2

Capillary (one cell thick)

Exchanging Materials

Exchanging Materials in Plants

Leaves are broad, thin and flat with lots of internal air spaces. This provides a large surface area, making them efficient at **photosynthesis**.

Leaves have stomata on their undersurface in order to…
- let **carbon dioxide in**
- let **oxygen out** (by diffusion)

(The exchange of substances is reversed during respiration.)

But photosynthesis also leads to loss of water vapour in a process called transpiration. Water loss is the price the plant must pay to photosynthesise. Transpiration is quicker in hot, dry, windy conditions.
- Water vapour from the internal leaf cells **evaporates** through the stomata.
- The size of the stomata is controlled by a pair of guard cells.

If plants lose water faster than it's taken up by the root hair cells, the stomata close to prevent wilting and **dehydration**.

During drought, photosynthesis might be impossible as the stomata close to prevent water loss.

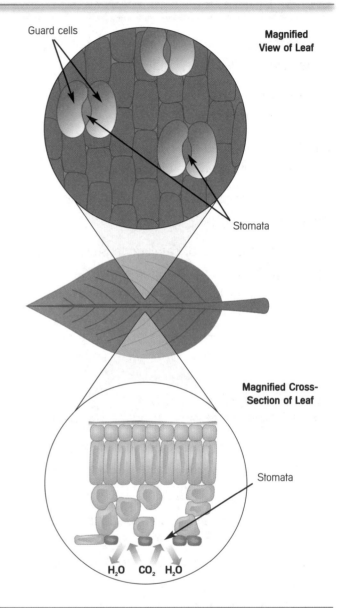

Guard cells

Magnified View of Leaf

Stomata

Magnified Cross-Section of Leaf

Stomata

H_2O CO_2 H_2O

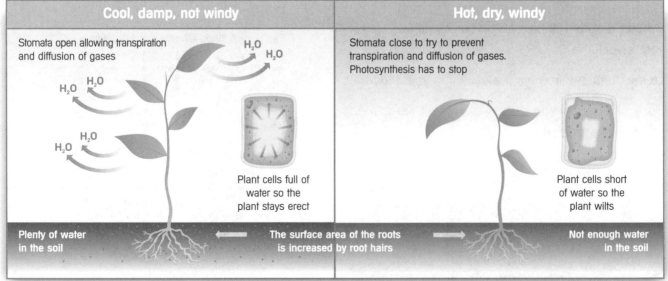

Cool, damp, not windy	Hot, dry, windy
Stomata open allowing transpiration and diffusion of gases	Stomata close to try to prevent transpiration and diffusion of gases. Photosynthesis has to stop
H_2O H_2O H_2O H_2O H_2O	
Plant cells full of water so the plant stays erect	Plant cells short of water so the plant wilts
Plenty of water in the soil	The surface area of the roots is increased by root hairs	Not enough water in the soil

The Circulation System

The **circulation system** carries blood from your heart to all the cells in your body. It consists of your **heart**, your **blood vessels** and your **blood**.

Blood is pumped to your lungs so carbon dioxide can be exchanged for oxygen. **Oxygenated blood** provides food and oxygen to cells. **Deoxygenated blood** takes away waste products (including carbon dioxide).

Blood flows around a 'figure of eight' circuit and passes through your heart twice on each circuit.
- Blood **travels away from** your heart through **arteries**.
- Blood **returns to** your heart through **veins**.

In your organs, blood flows through **capillaries**. Substances needed by the cells in your body tissues pass out of your blood, and substances produced by your cells pass into your blood through **capillary walls**.

There are two circulation systems:
1 To carry blood from your heart to your lungs then back to your heart.
2 To carry blood from your heart to all other organs then back to your heart.

- The **right side** of your heart pumps blood which is **low in oxygen** to your lungs, to pick up oxygen.
- The **left side** of your heart pumps blood which is **rich in oxygen** to all other parts of your body.

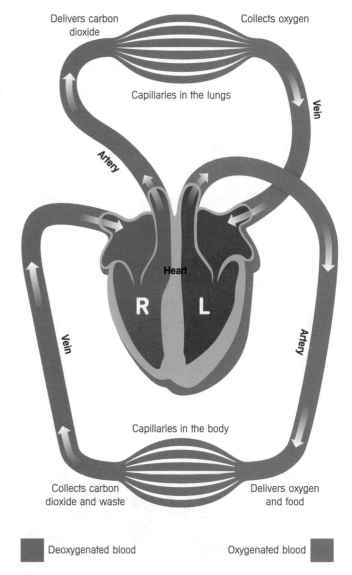

Delivers carbon dioxide

Collects oxygen

Capillaries in the lungs

Vein

Artery

Heart

R L

Vein

Artery

Capillaries in the body

Collects carbon dioxide and waste

Delivers oxygen and food

Deoxygenated blood   Oxygenated blood

The Blood

Blood has four components:
- plasma
- **red blood cells**
- white blood cells
- platelets.

Plasma is a straw-coloured liquid which transports…
- carbon dioxide from your organs to your lungs
- glucose from your small intestine to your organs
- other waste products (e.g. urea) from your liver to your kidneys.

Red blood cells transport oxygen from your lungs to your organs. They don't have a nucleus and contain lots of haemoglobin. Haemoglobin combines with oxygen in your lungs to form oxyhaemoglobin. In other organs, oxyhaemoglobin splits into haemoglobin and oxygen.

Key Words

Guard cells • Haemoglobin • Oxyhaemoglobin • Plasma • Root hair cells • Stomata • Transpiration • Wilting

Respiration

Aerobic Respiration

When **glucose** is combined with oxygen inside living cells it breaks down and releases **energy**. (The energy is contained inside the glucose molecule.)

This process is called **aerobic respiration**.

The energy released during aerobic respiration is used to make your muscles contract.

- Aerobic respiration occurs during normal day-to-day activity and provides for most of our energy needs.
- It doesn't produce energy as quickly as anaerobic respiration.
- It is a very efficient method of producing energy: one molecule of glucose produced by aerobic respiration can provide twenty times as much energy as anaerobic respiration.

A Working Muscle Cell

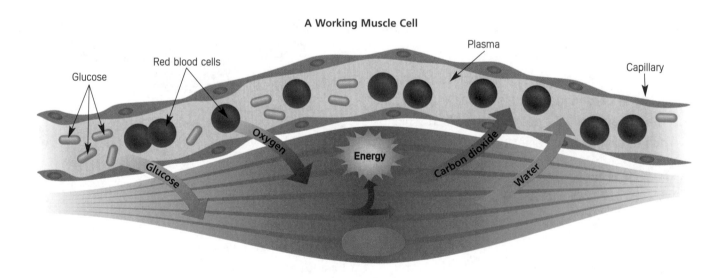

Glucose + Oxygen ⟶ Carbon dioxide	+	Water	+	Energy
Glucose and **oxygen** are brought to the respiring cells by the bloodstream.	**Carbon dioxide** is taken by the blood to the lungs, and breathed out.	**Water** passes into the blood and is lost as sweat, moist breath and urine.		**Energy** is used for muscle contraction, metabolism and maintaining temperature.

Anaerobic Respiration

If your muscles are subjected to long periods of vigorous activity, they become **fatigued** – they stop contracting efficiently, and hurt. If there isn't enough oxygen reaching your muscles, they use **anaerobic respiration** to obtain energy.

(HT) If no oxygen is present, glucose in living cells can't break down completely. Instead, a little energy is released very quickly inside your cells. This is **anaerobic respiration**.

The waste product from anaerobic respiration is **lactic acid** which accumulates in your tissues. When this happens, your muscles become fatigued.

After exercise, your body needs oxygen to break down the lactic acid; the oxygen needed is called an oxygen debt.

- Anaerobic respiration involves the **incomplete breakdown of glucose**. This means that much less energy is released than in aerobic respiration (about one twentieth).
- It can produce energy much faster than aerobic respiration over a short period of time.
- When the muscles are fatigued, deep breathing is required to oxidise the lactic acid to **carbon dioxide** and **water**.

| Glucose | ➡ | Energy | + | Lactic acid |

| Glucose from the bloodstream. | A small amount of energy is produced quickly and used for explosive activity. | Lactic acid accumulates in the muscles making them feel tired and 'rubbery'. |

Exercise and the Body

During exercise a number of changes take place:
- Your heart rate increases.
- The arteries supplying your muscles **dilate**.
- The rate and depth of your breathing increase.
- Blood flow to your muscles increases.
- The supply of oxygen and sugar is increased which speeds up removal of carbon dioxide.
- Animal starch, or **glycogen**, stored in your muscles is broken down to glucose to be used in respiration.

Key Words

Aerobic • Anaerobic • Dilate • Glycogen • Oxygen debt

The Kidneys and Dialysis

The Kidneys

Most people have two **kidneys**, situated on the back wall of the abdomen. Your kidneys **maintain the concentrations of dissolved substances** in your blood.

Your kidneys…

- regulate the amount of water in your blood
- regulate the amount of ions in your blood
- remove all **urea** in the form of **urine**.

If the kidneys fail, your body has no way of removing excess substances. This will ultimately result in death.

Each kidney is made up of two important tissues:

- **blood vessels**
- **tubules** (small tubes).

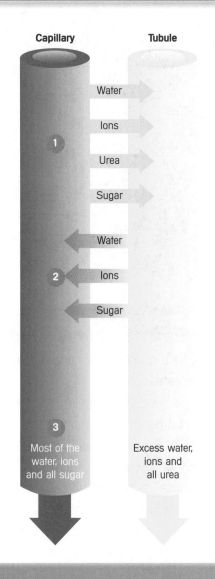

A kidney
There are millions of tiny tubules inside each kidney and each is very close to a blood capillary

Blood in

Blood out

Urine drains down ureter to bladder

How the Kidneys Function

Blood vessels take the blood through the kidney, where unwanted substances end up in millions of tiny tubules. These tiny tubules eventually join together to form the **ureter**. The substances flow through the tubules into the ureter, which leaves the kidney and ends up at the **bladder**.

So, there are three stages to learn.

1. **Ultrafiltration** – lots of water plus all the small molecules are squeezed out of the blood, under pressure, into the tubules.

2. **Selective Reabsorption** – the useful substances are reabsorbed into the blood from the tubules.

   (HT) Sugar and ions may have to be actively absorbed against a concentration gradient using energy (active transport).

3. **Excretion of Waste** – excess water, ions and all the **urea** now pass to the bladder in the form of **urine** and are eventually released from the body.

Capillary   Tubule

Water

Ions

1

Urea

Sugar

Water

2   Ions

Sugar

3

Most of the water, ions and all sugar

Excess water, ions and all urea

Key Words

Dialysis • Immune system • Irradiation • Partially permeable • Urea • Urine

Dialysis Machines

When a person's kidneys fail, they can use a **dialysis machine**.

As blood flows through a dialysis machine, it is separated from the dialysis fluid by **partially permeable membranes**. These membranes allow all the urea, and any excess substances, to pass from the blood to the dialysis fluid.

This restores the concentrations of dissolved substances in the blood to their normal levels.

Dialysis fluid contains the same concentration of useful substances as blood. This ensures **glucose** and **essential mineral ions** aren't lost through diffusion.

Dialysis must be carried out at regular intervals to maintain the patient's health.

Blood from patient

Blood returned to patient

All urea   Excess substances

Partially permeable membrane

Dialysis fluid

Waste fluid

Kidney Transplants

A kidney transplant allows a diseased kidney to be replaced by a healthy one from a **donor**. This is only performed if both kidneys fail (one kidney can do a good job).

The main problem with kidney transplants is the possibility of **rejection** by the **immune system**. So, precautions are taken to minimise the risk of rejection:

- A donor kidney with a **tissue type** as close as possible to that of the recipient is used. (This is best achieved if the donor is a close relative.)
- The bone marrow of the recipient is **irradiated** to stop the production of white blood cells.
- The recipient is treated with drugs which suppress the immune system.
- The recipient is kept in sterile conditions for some time after the operation to lessen the risk of infection due to their suppressed immune system.

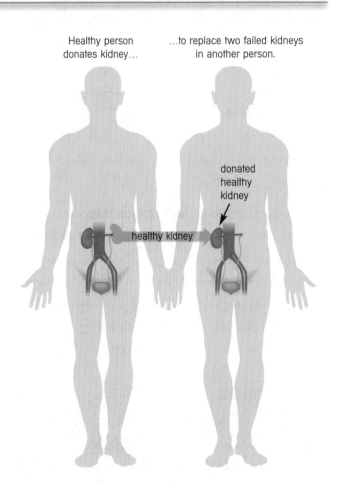

Healthy person donates kidney…   …to replace two failed kidneys in another person.

donated healthy kidney

healthy kidney

Using Microorganisms

Bacteria

Bacteria are used to make **yoghurt** and **cheese**.
Bacteria…
- vary in shape
- have a cell wall
- don't have a distinct nucleus.

Bacteria reproduce rapidly.

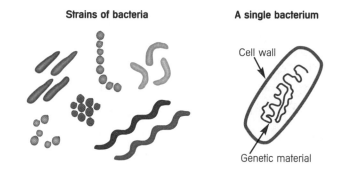

Strains of bacteria

A single bacterium

Cell wall

Genetic material

Yeast

Yeast is used to make **bread** and **alcoholic drinks**.

Yeast is a single-celled organism. Each yeast cell has…
- a nucleus
- cytoplasm
- a membrane
- a cell wall.

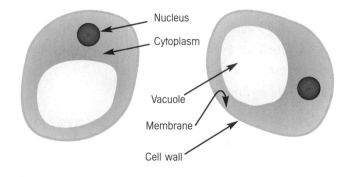

Nucleus

Cytoplasm

Vacuole

Membrane

Cell wall

How Yeast Works

Yeast can respire **without oxygen** (anaerobic respiration) to produce **ethanol** (alcohol) and **carbon dioxide**.

Glucose ⟶ Ethanol + Carbon dioxide + Energy

This is called **fermentation** and it has many industrial applications.

Yeast can also respire **with oxygen** (aerobic respiration) to produce **water** and **carbon dioxide**.

Glucose + Oxygen ⟶ Water + Carbon dioxide + Energy

Aerobic respiration produces more energy and is necessary for the yeast to grow and reproduce.

Key Words

Bacteria • Fermentation • Lactic acid • Malting • Yeast

Using Yeast in Baking

1. A mixture of **yeast** and **sugar** is added to flour.
2. The mixture is left in a warm place.
3. The **carbon dioxide** from the **respiring yeast** makes the dough rise.
4. The bubbles of gas in the dough expand when the bread is baked, making the bread 'light'.
5. As the bread is baked, any **alcohol** produced during respiration evaporates off.

Using Yeast in Brewing

1. In a process called malting, the starch in barley is broken down into a sugary solution by **enzymes**.
2. Yeast is added to the solution and **fermentation** takes place.
   - In beer-making, hops are added to give the beer flavour.
   - In wine-making, the yeast uses the natural sugars in the grapes as its energy source.
3. Carbon dioxide is bubbled off to leave the alcohol.

Using Bacteria to Make Yoghurt

1. A starter culture of **bacteria** is added to warm milk in a **fermenting vessel**.
2. The bacteria ferment the milk sugar (**lactose**) producing lactic acid which gives a sour taste.
3. The lactic acid causes the milk to clot and solidify into yoghurt.

The table below summarises these processes.

	Bread	Beer	Wine	Yoghurt
Microorganism	Yeast	Yeast	Yeast	Bacteria
Sugar Supply	Sugar added to flour	Starch in barley broken down into sugar (malting)	Grapes	Milk sugar (lactose)
Result	Carbon dioxide makes bread rise	Alcohol – hops added for flavour	Alcohol – flavour depends on grapes	Lactic acid clots and thickens milk

Growing Microorganisms

Growing Microorganisms

Microorganisms are grown in **fermenters** (large vessels).

Microorganisms grown in fermenters are used to make products such as **antibiotics** (e.g. penicillin) and mycoprotein (e.g. Quorn):

- **Penicillin** is made by growing penicillium, a **mould**, in a fermenter. The medium contains sugar and other nutrients which tend to be used for growth before the mould starts to make penicillin.
- **Mycoprotein**, a protein-rich food suitable for vegetarians, is made using Fusarium, a **fungus**. The fungus is grown on starch in aerobic conditions and the biomass is harvested and purified.

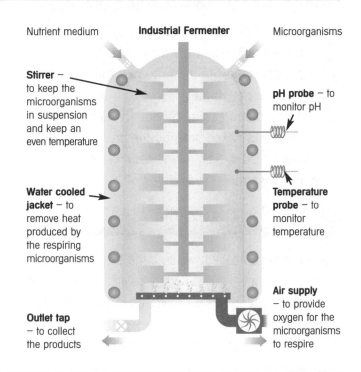

Nutrient medium **Industrial Fermenter** Microorganisms

Stirrer – to keep the microorganisms in suspension and keep an even temperature

pH probe – to monitor pH

Water cooled jacket – to remove heat produced by the respiring microorganisms

Temperature probe – to monitor temperature

Air supply – to provide oxygen for the microorganisms to respire

Outlet tap – to collect the products

Fuel Production

Fuels can be made from natural products by fermentation. But all oxygen must be excluded so anaerobic fermentation can occur. Biogas, which is mainly methane, can be produced in this way using a wide range of organic or waste materials containing **carbohydrates**.

Many different microorganisms are involved in the digestion of waste materials. Waste from sugar factories or sewage works can be used to provide biogas on a large scale. Biogas generators can supply the energy needs of individual families or farms on a small scale.

Anaerobic respiration can be used to produce **ethanol-based fuels** from sugar cane juices, or glucose derived from maize starch by the action of **carbohydrase** (an enzyme).

The ethanol produced needs to be distilled from the other products of fermentation, and can be used in motor vehicles.

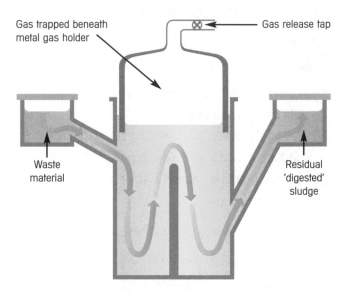

A Simple Biogas Generator

Gas trapped beneath metal gas holder

Gas release tap

Waste material

Residual 'digested' sludge

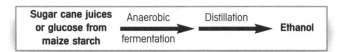

Sugar cane juices or glucose from maize starch	Anaerobic fermentation	Distillation	Ethanol

Key Words

Biogas • Culture medium • Fermentation • Incubation • Methane • Mycoprotein • Penicillin • Penicillium • Petri dish • Sterilised

Growing Microorganisms

Preparing a Culture Medium

Microorganisms are grown in a **culture medium** which contains **nutrients** that the particular microorganism might need, for example…

- carbohydrates (as an energy source)
- mineral ions
- vitamins
- proteins.

Agar is most commonly used as the growth medium. It is a soft, jelly-like substance (made from seaweed) which melts easily and re-solidifies at around 50°C.

Nutrients are added to the agar to provide ideal growing conditions for cultures.

Preparing Uncontaminated Cultures

If the cultures you want to investigate are **contaminated** by unwanted microorganisms, the 'rogue' microorganisms might produce undesirable substances which could be harmful.

It's only safe to use microorganisms if you have a pure culture of one species of microorganism.

To make useful products, uncontaminated cultures of microorganisms are prepared using the following procedures:

1 Sterilisation of Petri Dishes and Culture Medium
Both Petri dishes and culture medium are sterilised using an **autoclave** (a pressure cooker which exposes the dishes and the agar to high temperature and high pressure) to kill off unwanted microorganisms.

2 Sterilisation of Inoculating Loops
Inoculating loops tend to be made of nichrome wire with a wooden handle. They should be picked up like a pen, and the loop and half the wire should be heated to red heat in a Bunsen flame, then left to cool for five seconds. They are then sterile and can be used to transfer microorganisms to the culture medium. Don't blow on the loop or wave it around to cool it as it will pick up more microorganisms.

3 Sealing the Petri Dish
After the agar has been poured in and allowed to cool, the Petri dish should be sealed with tape (to prevent microorganisms from entering) and clearly labelled on the base. It should be stored upside down so **condensation** forms in the lid.

In schools and colleges, cultures are incubated at a maximum of 25°C to prevent the growth of potentially harmful **pathogens** that grow at body temperature (37°C).

In industry, higher temperatures can be used for more rapid growth.

Unit 3 Summary

Exchanging Materials

Water and dissolved substances move…
- along a concentration gradient
- from high to low concentrations
- by **osmosis** or **diffusion**.

HT **Active transport** = The process of substances absorbed against a concentration gradient using energy from respiration.

Exchanging Materials in Humans

Human organ systems are **specialised** to help the exchange of materials, for example…
- **villi** in the small intestine absorb products of digestion by diffusion and active transport
- **alveoli** in the lungs exchange oxygen and carbon dioxide (by diffusion).

Exchanging Materials in Plants

Leaves have stomata to let carbon dioxide in and oxygen out (by diffusion). This leads to **transpiration** (loss of water vapour).

The Circulation System

Circulation system = Heart, blood vessels and blood.

Oxygenated blood provides cells with **food** and **oxygen**.
Deoxygenated blood takes away **waste products**.

- Blood travels away from the heart through **arteries**.
- Blood returns to the heart through **veins**.
- In the organs, blood flows through **capillaries**.

There are two circulation systems:

1 Blood from heart ➡ Lungs ➡ Back to heart
2 Blood from heart ➡ All other organs ➡ Back to heart

Blood = Plasma, red blood cells, white blood cells and platelets.

Aerobic Respiration

Aerobic respiration = Respiration that uses oxygen.

Occurs during normal activity and produces more energy than anaerobic respiration (but less quickly).

Glucose + Oxygen ⟶ Carbon dioxide + Water (+ Energy)

Anaerobic Respiration

Anaerobic respiration = Respiration that doesn't use oxygen. Occurs during vigorous exercise.

HT Produces lactic acid, which needs oxygen to be broken down – **oxygen debt**.

Glucose ⟶ **Energy** + **Lactic acid**	

Kidneys and Dialysis

The kidneys maintain the concentrations of dissolved substances in the blood, and remove all **urea**.

Blood vessels take blood to kidneys ⟶ Unwanted substances enter tubules ⟶ Tubules join to form ureter ⟶ Unwanted substances reach bladder ⟶ Unwanted substances released

In a **dialysis machine**, blood flows between partially permeable membranes.

Kidney transplant replaces diseased kidney – rejection is a major problem, so…
* a donor kidney with a close tissue type is used (close relative)
* the patient's bone marrow is irradiated
* the patient's immune system is suppressed.

Microorganisms

Bacteria are used to make yoghurt and cheese.

Yeast is used to make bread and alcoholic drinks.

Fermentation = Glucose ⟶ Ethanol + Carbon dioxide + Energy

	Bread	Beer	Wine	Yoghurt
Microorganism	Yeast	Yeast	Yeast	Bacteria
Sugar Supply	Sugar is added to flour	Starch in barley is broken down into sugar (malting)	Grapes	Milk sugar (lactose)
Result	Released carbon dioxide makes bread rise	Alcohol – hops added to flavour	Alcohol – flavour depends on grapes	Lactic acid clots milk and thickens it

Microorganisms are grown in **fermenters**, and used to make products such as **penicillin** and **mycoprotein**. They are grown in a culture medium containing nutrients.

Uncontaminated cultures must be used. These can be prepared by…
* sterilisation of **Petri dishes** and culture medium (using an autoclave)
* sterilisation of inoculating loops (by heating in a Bunsen flame)
* sealing the Petri dish (sealed with tape and stored upside down).

Unit 3 Practice Questions

1 **a)** Name the process by which water can move in and out of a plant's root hair cell.

HT **b)** Explain the term 'active transport'.

c) How do the villi in the small intestine of an animal use active transport?

2 The diagram shows the human circulatory system.

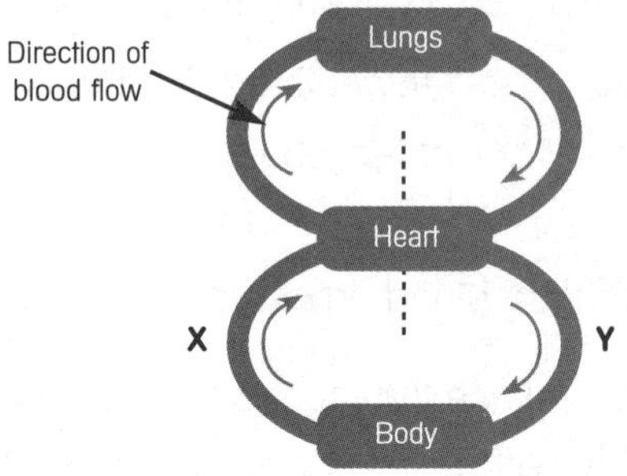

Direction of blood flow

Lungs

Heart

X   Y

Body

a) Name the type of blood vessel labelled X.

b) Name the type of blood vessel labelled Y.

c) Explain why this is called a double circulatory system.

3 Explain, as fully as you can, how the structure of a red blood cell is related to its function.

4 **a)** Write down the equation that describes aerobic respiration.

b) Match the words A, B, C and D with the spaces numbered 1 to 4 in the sentences below.

A oxygen debt

B anaerobic

C glucose

D lactic acid

__1__ respiration is the incomplete breakdown of __2__ into __3__ . Anaerobic respiration results in an __4__ .

5 The function of the kidney is to maintain the concentration of dissolved substances in the blood and remove all harmful substances. Name two substances that are removed by the kidneys.

i) .... ii) ....

6 Match the words A, B, C and D with the parts labelled 1 to 4 in the diagram below.

A partially permeable membrane .... **B** urea ....

C blood from patient .... **D** waste fluid ....

7 Match each of the words A, B, C and D with the name of the product it is used to make.

A bacteria .... **B** yeast ....

C penicillium .... **D** fusarium ....

1 mycoprotein **2** antibiotics
3 bread **4** yoghurt

8 Name the large vessel that microorganisms are grown in.

....

9 a) When preparing a culture medium, what nutrients should you include?

....

....

b) How would you sterilise an inoculating loop?

....

c) i) At what temperature would a school or college incubate a Petri dish of microorganisms?

....

ii) Why do they use this temperature?

....

....

Glossary of Key Words

Adaptation − the gradual change of a particular organism over generations to become better suited to its environment.

Aerobic − with oxygen.

Aerobic respiration − respiration which uses oxygen.

Allele − an alternative form of a particular gene.

Alveoli − air sacs in the lungs. Oxygen diffuses into them and carbon dioxide diffuses out of them.

Amylase − an enzyme that breaks down starch.

Anaerobic − without oxygen.

Antibiotics − medication used to kill bacterial pathogens inside the body.

Asexual reproduction − when new individuals are produced which are identical to the parents; doesn't involve the fusion of gametes.

Bacteria − a single-celled microorganism that has no nucleus.

Bile − a greenish-yellow fluid produced by the liver.

Biodiversity − the variety among living organisms and the ecosystems in which they live.

Biogas − fuel produced from the anaerobic decomposition of organic waste.

Biomass − the mass of a plant or animal without the water content.

Carbon cycle − the constant recycling of carbon by the processes in life, death and decay.

Carcinogen − a substance that causes cancer.

Catalyst − a substance that increases the rate of a chemical reaction without being changed itself.

Cell − a fundamental unit of a living organism.

Cholesterol − a fatty substance which is found in all cells of the body.

Chlorophyll − the green pigment found in most plants; responsible for photosynthesis.

Chloroplast − tiny structure in the cytoplasm of plant cells which contains chlorophyll.

Chromosomes − long molecules found in the nucleus of all cells containing DNA.

Clone − a genetically identical offspring of an organism.

Community − all the living organisms in an area or habitat.

Concentration gradient − a change in the concentration of a substance from one region to another.

Culture medium − a nutrient system used for the artificial growth of bacteria and other cells.

Cytoplasm − the substance found in living cells (outside the nucleus) where chemical reactions take place.

Decay − to rot or decompose.

Deficiency disease − a disease caused by the lack of an essential element in the diet.

Deforestation − the destruction of forests by cutting down trees.

Detritus − organic material formed from dead and decomposing plants and animals.

Diabetes − a condition in which a person's blood glucose level rises to very high concentrations.

Dialysis − the artificial removal of urea and excess material from the blood. (Used when the kidneys fail.)

Differentiation − making / becoming different.

Diffusion − the mixing of two substances through the natural movement of their particles from a high concentration to a low concentration.

Dilate − to widen or enlarge.

DNA − nucleic acid molecules which contain genetic information and make up chromosomes.

Drug – a chemical substance that alters the way the body works.

Effector – the part of the body, e.g. a muscle or a gland, which produces a response to a sensor.

Embryo – a ball of cells which will develop into a human / animal baby.

Environment – the conditions around an organism.

Enzyme – a protein which speeds up a reaction (a 'biological catalyst').

Evolve – to change naturally over a period of time.

Exponentially – with accelerating speed.

Extinct – a species that has died out.

Fermentation – the conversion of sugar to alcohol and carbon dioxide using yeast.

Fertilisation – the fusion of the male gamete with the female gamete.

Fetus – an unborn human / animal baby.

Food chain – the feeding relationship between organisms.

Fossil – the remains of animals / plants preserved in rock.

FSH (Follicle Stimulating Hormone) – a hormone that stimulates ovaries to produce oestrogen.

Gamete – a specialised sex cell formed by meiosis.

Gene – part of a chromosome, made up of DNA; controls a certain characteristic.

Gland – an organ in an animal body that secretes substances.

Global warming – the increase in the average temperature on Earth due to a rise in the levels of greenhouse gases in the atmosphere. .

Glycogen – a form of starch in which sugars are stored in the body for energy.

Greenhouse effect – the process by which the Earth is kept warm by the ozone reflecting heat back to Earth.

Guard cells – pairs of sausage-shaped plant cells which open and close to allow oxygen into the leaf and water and carbon dioxide out (through the stomata).

Haemoglobin – red pigment in the red blood cells which carries oxygen to the organs.

Herbicide – a toxic substance used to destroy unwanted vegetation.

Hormone – a regulatory substance which stimulates cells or tissues into action; produced by a gland.

Immune system – the body's defence system against infections and diseases (consists of white blood cells and antibodies).

Incubation – growing in a laboratory under controlled conditions.

Infectious – a disease that is easily spread through air, water, etc.

Insulin – a hormone, produced by the pancreas, which controls blood glucose concentrations.

Ion (electrolyte) – a mineral that the body needs; a particle that has a positive or negative electrical charge.

Irradiation – exposure to radiation to kill microorganisms.

Leprosy – a contagious bacterial disease affecting the skin and nerves.

LH (Luteinising Hormone) – a hormone that stimulates changes in the menstrual cycle.

Lipase – an enzyme which breaks down fat into fatty acids and glycerol.

Lipoproteins – structured materials of the cell membrane made up of lipid (fat) and protein joined together.

Glossary of Key Words

Malnourished – suffering from a lack of essential food nutrients.

Malting – e.g. germinating barley under controlled conditions then drying it in a kiln (oven) – the starch is converted to glucose.

Menstrual cycle – the monthly cycle of hormonal changes in a woman.

Metabolic rate – the rate at which an animal uses energy over a given time period.

Methane – clear gas given off by animal waste; can be used as a fuel; a greenhouse gas.

Mineral ions – charged particles formed from elements (or groups of elements) which plants need for healthy growth.

Mitochondria – the structure in the cytoplasm where energy is produced from the chemical reactions.

Mitosis – cell division that forms two daughter cells, each with the same number of chromosomes as the parent cell.

MRSA (Methicillin-resistant Staphylococcus Aureus) – an antibiotic-resistant bacterium; a 'superbug'.

Mutation – a change in the genetic material of a cell.

Mycoprotein – a protein-rich food produced from fungi.

Natural selection – the survival of individual organisms that are best suited / adapted to their environment.

Neurone – a specialised cell which transmits electrical messages or nerve impulses.

Nitrate – any compound containing the nitrate radical (NO_3).

Non-renewable – energy sources that cannot be replaced in a lifetime.

Nucleus – the control centre of a cell.

Obesity – the condition of being very overweight.

Osmosis – the movement of water through a partially permeable membrane into a solution with lower water concentration.

Oxyhaemoglobin – the combination of oxygen and haemoglobin.

Partially permeable – a barrier which allows only certain substances through.

Pathogen – a disease-causing microorganism.

Penicillin – an antibiotic drug used to treat bacterial infection (discovered by Alexander Fleming).

Penicillium – a mould, from which penicillin is developed.

Pesticide – a substance used for destroying insects or other pests.

Petri dish – a round, shallow dish used to grow bacteria.

Photosynthesis – the chemical process that uses light energy to produce glucose in green plants.

Pituitary gland – a small gland at the base of the brain that produces hormones.

Plasma – the clear fluid part of blood that contains proteins and minerals.

Pollution – the contamination of an environment by chemicals, waste or heat.

Predator – an animal that hunts, kills and eats its prey.

Protease – an enzyme used to break down proteins into amino acids.

Radiation – electromagnetic particles / rays emitted by a radioactive substance.

Receptor – a sense organ, e.g. eyes, ears, nose.

Reflex action – an involuntary action, e.g. automatically jerking your hand away from something hot.

Respiration – the process of converting glucose into energy inside cells.

Ribosomes – small structures found in the cytoplasm of living cells, where protein synthesis takes place.

Root hair cells – found on the roots of plants; absorb water from the soil.

Saturated fat – animal fat, containing no double carbon carbon bonds; considered to be unhealthy.

Sexual reproduction – when new individuals are produced which are not genetically identical to the parents; involves the fusion of gametes.

Side-effect – condition caused by taking medication, e.g. headache, nausea.

Specialised – adapted for a particular purpose.

Stem cell – a cell of human embryos or adult bone marrow which has yet to differentiate.

Sterilised – free from all microorganisms.

Stomata – openings / pores in leaves.

Surface area – the external area of a living thing.

Sustainable – resources that can be replaced.

Synapse – the gap between two neurones.

Thermoregulation – the maintenance of a constant body temperature in warm-blooded animals.

Toxin – a poison produced by a living organism.

Transpiration – the evaporation of water from plants (through the stomata).

Unsaturated fat – vegetable fat; considered to be healthy.

Urea – waste product of proteins formed in the liver and excreted in urine.

Urine – water and waste products filtered by the kidneys.

Vaccine – a liquid preparation used to make the body produce antibodies to provide protection against disease.

Vacuole – a fluid-filled sac found in cytoplasm.

Variation – differences between individuals of the same species.

Villi – projections which stick out from the walls of the small intestine. Each villus contains a network of blood capillaries for absorbing soluble food.

Wilting – the drooping of a plant caused by excessive water loss through transpiration.

Yeast – a single-celled fungus; a microorganism.

Zygote – a cell formed by the fusion of the nuclei of a male sex cell and a female sex cell (gametes).

(HT) **Active transport** – the movement of substances against a concentration gradient; requires energy.

Lactic acid – a compound produced when cells respire without oxygen (i.e. anaerobically).

Meiosis – cell division that forms daughter cells with half the number of chromosomes of the parent cell.

Oxygen debt – oxygen deficiency caused by intense / vigorous exercise.

Answers to Practice Questions

Unit 1a

1. A 3; B 2; C 4; D 1.
2. Synapse.
3. A 2; B 4; C 3; D 1.
4. a) FSH is released from the pituitary gland. It acts on the ovaries, stimulating the egg to mature and be released.
   b) Oestrogen.
   c) It inhibits the production of FSH. Eggs don't mature and are not released.
   d) LH (luteinising hormone).
5. A 2; B 4; C 1; D 3.

6. The rate at which all chemical reactions in the cells of the body are carried out.
7. A 3; B 2; C 1; D 4.
8. A 2; B 4; C 1; D 3.
9. a) iv) Dead pathogen injected – Antibodies produced – White blood cells sensitised – Immunity.
   b) iv) Antibiotics can be used to fight infections caused by bacteria.
   c) iii) White blood cells produce antibiotics.
10. Because many strains of bacteria have developed resistance to antibiotics as a result of natural selection.

Unit 1b

1. A 3; B 1; C 4; D 2.
2. **Any two from:** Food; Light; Space; Water; Nutrients; Mates; Territory.
3. A 3; B 1; C 2; D 4.
4. a) iv) Sexual reproduction.
   b) iii) The offspring are identical to each other but not the parents.
   c) iii) Sexual reproduction.
5. Differences between individuals of the same species.
6. An adult plant is selected. Some cells are scraped off and are added to soil containing nutrients and hormones. (The new plants produced are identical to the parent plant.)

7. **Any three from:** To improve crop yield; To improve resistance to pests / herbicides; To extend shelf-life; To harness the cell chemistry of an organism so that it produces a required substance.
8. a) The idea that all living things evolved from simple life forms that first developed three billion years ago.
   b) The remains of plants or animals from many years ago which are found in rock.
9. A 2; B 3; C 4; D 1 **or** A 2; B 4; C 3; D 1.
10. A 3; B 1; C 4; D 2.
11. Economic development, social development and environmental protection.

Unit 2

1. a) **A** = cell membrane; **B** = nucleus; **C** = ribosome.
   b) This is where most energy is released during respiration.
   c) Cell membrane.
2. a) Light; Carbon dioxide; Water; Chlorophyll.
   b) Glucose; Oxygen.
   c) **i)** Diffusion
      ii) Osmosis
   d) Yellow leaves.
3. **A** = blackbirds; **B** = ladybirds; **C** = greenfly; **D** = oak tree.

4. a) ii) The pancreas
   b) iii) It could rise to a fatally high level.
5. a) **Any two from:** Respiration; Protein synthesis; Photosynthesis.
   b) Temperature; pH.
   c) Bile neutralises stomach acid and emulsifies fats.
6. a) iii) DNA
   b) ii) 23
   c) iv) 100%

Unit 3

1. a) Osmosis.
   b) Active transport is the movement of substances in and out of cells against a concentration gradient.
   c) Villi line the walls of the small intestine and, because they have a massive surface area and large network of capillaries, they can absorb the products of digestion by diffusion and active transport.
2. a) **X** = Vein.
   b) **Y** = Artery.
   c) There are two separate circulation systems: one carrying blood from the heart to the lungs and back to the heart, and the other carrying blood to all the other organs of the body and then back to the heart.

3. It has no nucleus so it can contain more haemoglobin and it can carry more oxygen which can combine to make oxyhaemoglobin.
4. a) Glucose + Oxygen → Carbon dioxide + Water + Energy
   b) A 4; B 1; C 2; D 3.
5. **Any two from:** Urea; Excess water; Excess ions.
6. A 3; B 2; C 1; D 4.
7. A 4; B 3; C 2; D 1.
8. Fermenter.
9. a) Carbohydrates, mineral ions, vitamins and proteins.
   b) Place the loop in a Bunsen flame and heat to red heat.
   c) i) 25°C.
      ii) To prevent the growth of pathogens that would grow at body temperature.

Notes

Index